INVESTIGATIONS G

Pebbles, Sand, and Silt

Full Option Science System
Developed at the Lawrence Hall of Science, University of California, Berkeley
Published and Distributed by Delta Education

FOSS Lawrence Hall of Science Team
Larry Malone and Linda De Lucchi, FOSS Project Codirectors and Lead Developers
Kathy Long, FOSS Assessment Director; David Lippman, Program Manager; Carol Sevilla, Publications Design Coordinator; Susan Stanley, Illustrator; John Quick, Photographer
FOSS Curriculum Developers: Brian Campbell, Teri Lawson, Alan Gould, Susan Kaschner Jagoda, Ann Moriarty, Jessica Penchos, Kimi Hosoume, Virginia Reid, Joanna Snyder, Erica Beck Spencer, Joanna Totino, Diana Velez, Natalie Yakushiji
Susan Ketchner, Technology Project Manager
FOSS Technology Team: Dan Bluestein, Christopher Cianciarulo, Matthew Jacoby, Kate Jordan, Frank Kusiak, Nicole Medina, Jonathan Segal, Dave Stapley, Shan Tsai

Delta Education Team
Bonnie A. Piotrowski, Editorial Director, Elementary Science
Project Team: Jennifer Apt, Sandra Burke, Joann Hoy, Kristen Mahoney, Jennifer McKenna

Thank you to all FOSS Grades K-6 Trial Teachers
Heather Ballard, Wilson Elementary, Coppell, TX; Mirith Ballestas De Barroso, Treasure Forest Elementary, Houston, TX; Terra L. Barton, Harry McKillop Elementary, Melissa, TX; Rhonda Bernard, Frances E. Norton Elementary, Allen, TX; Theresa Bissonnette, East Millbrook Magnet Middle School, Raleigh, NC; Peter Blackstone, Hall Elementary School, Portland, ME; Tiffani Brisco, Seven Hills Elementary, Newark, TX; Darrow Brown, Lake Myra Elementary School, Wendell, NC; Heather Callaghan, Olive Chapel Elementary, Apex, NC; Katie Cannon, Las Colinas Elementary, Irving, TX; Elaine M. Cansler, Brassfi eld Road Elementary School, Raleigh, NC; Kristy Cash, Wilson Elementary, Coppell, TX; Monica Coles, Swift Creek Elementary School, Raleigh, NC; Shirley Conner, Ocean Avenue Elementary School, Portland, ME; Sally Connolly, Cape Elizabeth Middle School, Cape Elizabeth, ME; Melissa Cook-Airhart, Harry McKillop Elementary, Melissa, TX; Melissa Costa, Olive Chapel Elementary, Apex, NC; Hillary P. Croissant, Harry McKillop Elementary, Melissa, TX; Rene Custeau, Hall Elementary School, Portland, ME; Nancy Davis, Martha and Josh Morriss Mathematics and Engineering Elementary School, Texarkana, TX; Nancy Deveneau, Wilson Elementary, Coppell, TX; Karen Diaz, Las Colinas Elementary, Irving, TX; Marlana Dumas, Las Colinas Elementary, Irving, TX; Mary Evans, R.E. Good Elementary School, Carrollton, TX; Jacquelyn Farley, Moss Haven Elementary, Dallas, TX; Corinna Ferrier, Oak Forest Elementary, Humble, TX; Allison Fike, Wilson Elementary, Coppell, TX; Barbara Fugitt, Martha and Josh Morriss Mathematics and Engineering Elementary School, Texarkana, TX; Colleen Garvey, Farmington Woods Elementary, Cary, NC; Judy Geller, Bentley Elementary School, Oakland, CA; Erin Gibson, Las Colinas Elementary, Irving, TX; Kelli Gobel, Melissa Ridge Intermediate School, Melissa, TX; Dollie Green, Melissa Ridge Intermediate School, Melissa, TX; Brenda Lee Harrigan, Bentley Elementary School, Oakland, CA; Cori Harris, Samuel Beck Elementary, Trophy Club, TX; Kim Hayes, Martha and Josh Morriss Mathematics and Engineering Elementary School, Texarkana, TX; Staci Lynn Hester, Lacy Elementary School, Raleigh, NC; Amanda Hill, Las Colinas Elementary, Irving, TX; Margaret Hillman, Ocean Avenue Elementary School, Portland, ME; Cindy Holder, Oak Forest Elementary, Humble, TX; Sarah Huber, Hodge Road Elementary, Knightdale, NC; Susan Jacobs, Granger Elementary, Keller, TX; Carol Kellum, Wallace Elementary, Dallas, TX; Jennifer A. Kelly, Hall Elementary School, Portland, ME; Brittani Kern, Fox Road Elementary, Raleigh, NC; Jodi Lay, Lufkin Road Middle School, Apex, NC; Melissa Lourenco, Lake Myra Elementary School, Wendell, NC; Ana Martinez, RISD Academy, Dallas, TX; Shaheen Mavani, Las Colinas Elementary, Irving, TX; Mary Linley McClendon, Math Science Technology Magnet School, Richardson, TX; Adam McKay, Davis Drive Elementary, Cary, NC; Leslie Meadows, Lake Myra Elementary School, Wendell, NC; Anne Mechler, J. Erik Jonsson Community School, Dallas, TX; Anne Miller, J. Erik Jonsson Community School, Dallas, TX; Shirley Diann Miller, The Rice School, Houston, TX; Keri Minier, Las Colinas Elementary, Irving, TX; Stephanie Renee Nance, T.H. Rogers Elementary, Houston, TX; Cynthia Nilsen, Peaks Island School, Peaks Island, ME; Elizabeth Noble, Las Colinas Elementary, Irving, TX; Courtney Noonan, Shadow Oaks Elementary School, Houston, TX; Sarah Peden, Aversboro Elementary School, Garner, NC; Carrie Prince, School at St. George Place, Houston, TX; Marlaina Pritchard, Melissa Ridge Intermediate School, Melissa, TX; Alice Pujol, J. Erik Jonsson Community School, Dallas, TX; Claire Ramsbotham, Cape Elizabeth Middle School, Cape Elizabeth, ME; Paul Rendon, Bentley Elementary, Oakland, CA; Janette Ridley, W.H. Wilson Elementary School, Coppell, TX; Kristina (Crickett) Roberts, W.H. Wilson Elementary School, Coppell, TX; Heather Rogers, Wendell Creative Arts & Science Magnet Elementary School, Wendell, NC; Alissa Royal, Melissa Ridge Intermediate School, Melissa, TX; Megan Runion, Olive Chapel Elementary, Apex, NC; Christy Scheef, J. Erik Jonsson Community School, Dallas, TX; Samrawit Shawl, T.H. Rogers School, Houston, TX; Nicole Spivey, Lake Myra Elementary School, Wendell, NC; Ashley Stephenson, J. Erik Jonsson Community School, Dallas, TX; Jolanta Stern, Browning Elementary School, Houston, TX; Gale Stimson, Bentley Elementary, Oakland, CA; Ted Stoeckley, Hall Middle School, Larkspur, CA; Cathryn Sutton, Wilson Elementary, Coppell, TX; Camille Swander, Ocean Avenue Elementary School, Portland, ME; Brandi Swann, Westlawn Elementary School, Texarkana, TX; Robin Taylor, Arapaho Classical Magnet, Richardson, TX; Michael C. Thomas, Forest Lane Academy, Dallas, TX; Jomarga Thompkins, Lockhart Elementary, Houston, TX; Mary Timar, Madera Elementary, Lake Forest, CA; Helena Tongkeamha, White Rock Elementary, Dallas, TX; Linda Trampe, J. Erik Jonsson Community School, Dallas, TX; Charity VanHorn, Fred A. Olds Elementary, Raleigh, NC; Kathleen VanKeuren, Lufkin Road Middle School, Apex, NC; Valerie Vassar, Hall Elementary School, Portland, ME; Megan Veron, Westwood Elementary School, Houston, TX; Mary Margaret Waters, Frances E. Norton Elementary, Allen, TX; Stephanie Robledo Watson, Ridgecrest Elementary School, Houston, TX; Lisa Webb, Madisonville Intermediate, Madisonville, TX; Matt Whaley, Cape Elizabeth Middle School, Cape Elizabeth, ME; Nancy White, Canyon Creek Elementary, Austin, TX; Barbara Yurick, Oak Forest Elementary, Humble, TX; Linda Zittel, Mira Vista Elementary, Richmond, CA

Photo Credits: © Tom Gowanlock/Shutterstock (cover); © Francesca/Dreamstime; © Laurie Meyer; © Erica Beck Spencer

Published and Distributed by Delta Education, a member of the School Specialty Family
The FOSS program was developed in part with the support of the National Science Foundation grant nos. MDR-8751727 and MDR-9150097. However, any opinions, findings, conclusions, statements, and recommendations expressed herein are those of the authors and do not necessarily reflect the views of NSF. FOSSmap was developed in collaboration between the BEAR Center at UC Berkeley and FOSS at the Lawrence Hall of Science.

Pebbles, Sand, and Silt
Investigations Guide, 1487590
978-1-62571-295-0
Printing 5 – 6/2017
Webcrafters, Madison, WI

Pebbles, Sand, and Silt

TABLE OF CONTENTS

Welcome to FOSS® Next Generation™

Getting Started with FOSS Next Generation for Grades K–2

Whether you're new to hands-on science or a FOSS veteran, you'll be up and running in no time and ready to lead your students on a fantastic voyage through the wonders of the natural and designed world.

Watch our short video series or browse the next few pages to get started!

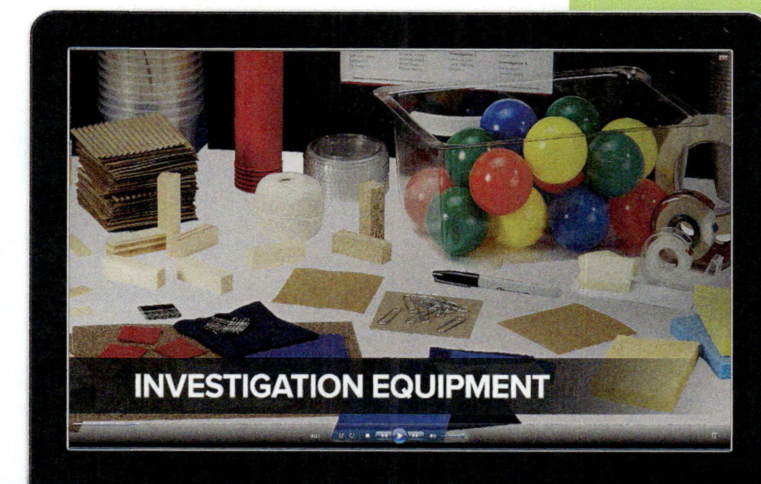

Getting Started with FOSS: Meet Your Module video

Scan here or visit deltaeducation.com/goFOSS

Three-Dimensional Active Science

It's time to experience the three dimensions of the NGSS—**disciplinary core ideas**, **crosscutting concepts**, and **science and engineering practices**. Engage in rich investigations that immerse your students in real-world applications of important scientific phenomenon, supported by just-in-time teaching tips and strategies.

Getting Started with Your Equipment Kit

Meet Your FOSS Module!

Your FOSS module includes one or more large boxes, called drawers, and two smaller boxes for the Teacher Toolkit, student books, and other equipment. Each drawer has a label on the front listing its contents. Your packing list is always in Drawer 1.

Permanent Equipment

Your equipment kit includes enough permanent equipment for up to 8 groups (32 students). This equipment is classroom-tested and expected to last 7–10 years.

Consumable Equipment

Your kit also includes consumable materials for three class uses. Convenient refill kits provide materials for three additional uses and are available through Delta Education.

Easy Set-up and Clean-up!

FOSS Next Generation equipment drawers are packed by investigation to facilitate prep and to make packing up for the next use a snap!

Drawer sections include:

- Unique materials needed for one investigation
- Common equipment used in multiple investigations
- Consumable materials—when it's empty you know it's time to refill!

Order Refills Online

deltaeducation.com/ refillcenter

Live Organisms

Some investigations require live organisms. Schools are encouraged to purchase these organisms from a local biological supply company to minimize both transit time and the impact of adverse weather on the health of the organisms.

If living material cards are purchased from Delta Education, they will be shipped separately in a green and white envelope. Keep these cards in a safe place until it's time to redeem them for the investigation.

Call Delta Education at 800-258-1302 at least three weeks before you need your organisms.

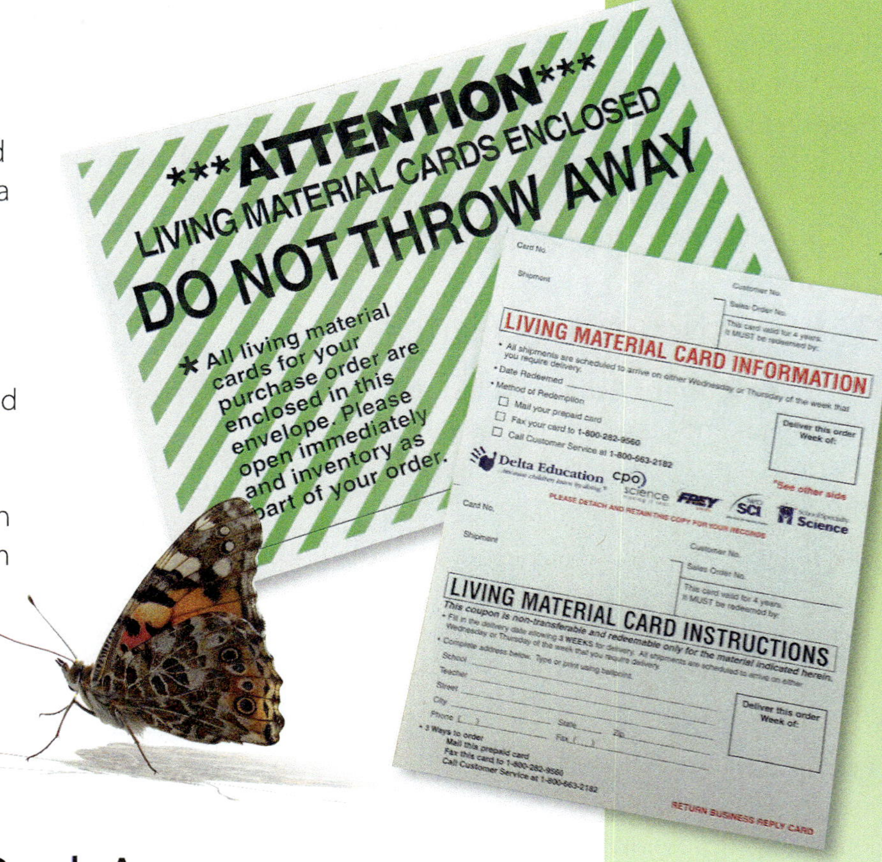

Premium Student eBook Access

If your school purchased a premium class license for the *FOSS Science Resources* student eBook, your access codes will be shipped separately in a blue and white striped envelope. Use this access code on FOSSweb to unlock student eBook access.

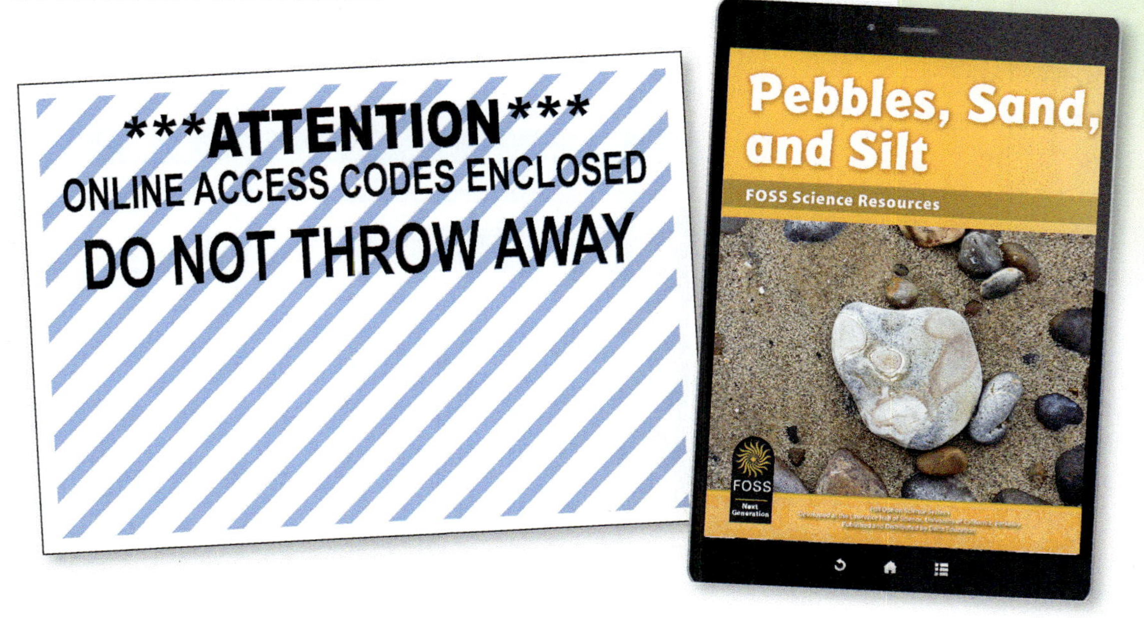

Getting Started with Your Teacher Toolkit

The Teacher Toolkit is the most important part of the FOSS program. There are three parts of the Teacher Toolkit—the ***Investigations Guide***, ***Teacher Resources***, and the student ***Science Resources* Book.** It's here that all the wisdom and experience from years of research and classroom development comes together to support teachers with lesson facilitation and in-depth strategies for taking investigations to the next level.

Pebbles, Sand, and Silt
INVESTIGATIONS GUIDE

1. Investigations Guide

The ***Investigations Guide*** is your roadmap to prepare for and lead the FOSS investigations. Chapters are tabbed for easy access to important module information.

The module **Overview** gives you a high-level look at the 10–12 weeks of instruction in each module including a summary matrix, schedule for the module, and product support contacts.

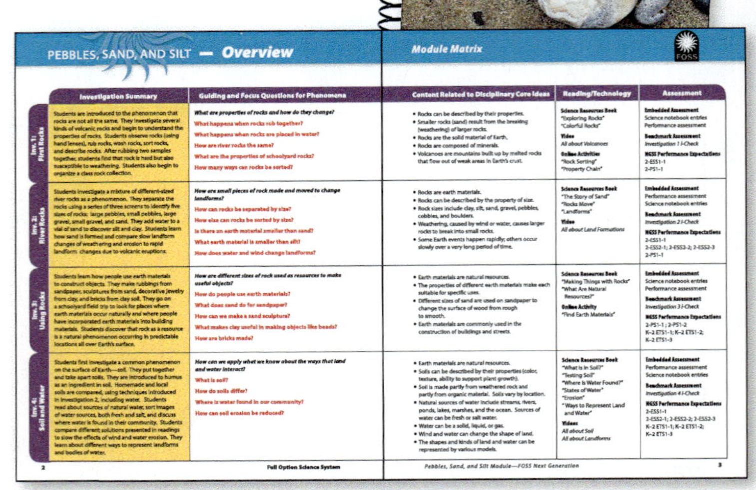

PEBBLES, SAND, AND SILT — *Overview* | *Module Matrix*

	Investigation Summary	Guiding and Focus Questions for Phenomena	Content Related to Disciplinary Core Ideas	Reading/Technology	Assessment
Inv. 1: First Rocks	Students are introduced to the phenomenon that rocks are not all the same. They investigate several kinds of volcanic rocks and begin to understand the properties of rocks. Students observe rocks (using hand lenses), rub rocks, wash rocks, sort rocks, and describe rocks. After rubbing two samples together, students find that rock is hard but also susceptible to weathering. Students also begin to organize a class rock collection.	*What are properties of rocks and how do they change?* What happens when rocks rub together? What happens when rocks are placed in water? How are river rocks the same? What are the properties of schoolyard rocks? How many ways can rocks be sorted?	• Rocks can be described by their properties. • Smaller rocks (sand) result from the breaking (weathering) of larger rocks. • Rocks are the solid material of Earth. • Rocks are composed of minerals. • Volcanoes are mountains built up by melted rocks that flow out of weak areas in Earth's crust.	**Science Resources Book** "Exploring Rocks" "Colorful Rocks" **Video** *All about Volcanoes* **Online Activities** "Rock Sorting" "Property Chain"	**Embedded Assessment** Science notebook entries Performance assessment **Benchmark Assessment** *Investigation 1 I-Check* **NGSS Performance Expectations** 2-ESS1-1 2-PS1-1
Inv. 2: River Rocks	Students investigate a mixture of different-sized river rocks as a phenomenon. They separate the rocks using a series of three screens to identify five sizes of rocks: large pebbles, small pebbles, large gravel, small gravel, and sand. They add water to a vial of sand to discover silt and clay. Students learn how sand is formed and compare slow landform changes of weathering and erosion to rapid landform changes due to volcanic eruptions.	*How are small pieces of rock made and moved to change landforms?* How can rocks be separated by size? How else can rocks be sorted by size? Is there an earth material smaller than sand? What earth material is smaller than silt? How does water and wind change landforms?	• Rocks are earth materials. • Rocks can be described by the property of size. • Rock sizes include clay, silt, sand, gravel, pebbles, cobbles, and boulders. • Weathering, caused by wind or water, causes larger rocks to break into small rocks. • Some Earth events happen rapidly; others occur slowly over a very long period of time.	**Science Resources Book** "The Story of Sand" "Rocks Move" "Landforms" **Video** *All about Land Formations*	**Embedded Assessment** Performance assessment Science notebook entries **Benchmark Assessment** *Investigation 2 I-Check* **NGSS Performance Expectations** 2-ESS1-1 2-ESS2-1; 2-ESS2-2; 2-ESS2-3 2-PS1-1
Inv. 3: Using Rocks	Students learn how people use earth materials to construct objects. They make rubbings from sandpaper, sculptures from sand, decorative jewelry from clay, and bricks from clay soil. They go on a schoolyard field trip to look for places where earth materials occur naturally and where people have incorporated earth materials into building materials. Students discover that rock as a resource is a natural phenomenon occurring in predictable locations all over Earth's surface.	*How are different sizes of rock used as resources to make useful objects?* How do people use earth materials? What does sand do for sandpaper? How can we make a sand sculpture? What makes clay useful in making objects like beads? How are bricks made?	• Earth materials are natural resources. • The properties of different earth materials make each suitable for specific uses. • Different sizes of sand are used on sandpaper to change the surface of wood from rough to smooth. • Earth materials are commonly used in the construction of buildings and streets.	**Science Resources Book** "Making Things with Rocks" "What Are Natural Resources?" **Online Activity** "Find Earth Materials"	**Embedded Assessment** Science notebook entries Performance assessment **Benchmark Assessment** *Investigation 3 I-Check* **NGSS Performance Expectations** 2-PS1-1 ; 2-PS1-2 K–2 ETS1-1; K–2 ETS1-2; K–2 ETS1-3
Inv. 4: Soil and Water	Students first investigate a common phenomenon on the surface of Earth—soil. They put together and take apart soils. They are introduced to humus as an ingredient in soil. Homemade and local soils are compared, using techniques introduced in Investigation 2, including water. Students read about sources of natural water, sort images of water sources, both fresh and salt, and discuss where water is found in their community. Students compare different solutions presented in readings to slow the effects of wind and water erosion. They learn about different ways to represent landforms and bodies of water.	*How can we apply what we know about the ways that land and water interact?* What is soil? How do soils differ? Where is water found in our community? How can soil erosion be reduced?	• Earth materials are natural resources. • Soils can be described by their properties (color, texture, ability to support plant growth). • Soil is made partly from weathered rock and partly from organic material. Soils vary by location. • Natural sources of water include streams, rivers, ponds, lakes, marshes, and the ocean. Sources of water can be fresh or salt water. • Water can be a solid, liquid, or gas. • Wind and water can change the shape of land. • The shapes and kinds of land and water can be represented by various models.	**Science Resources Book** "What Is in Soil?" "Testing Soil" "Where Is Water Found?" "States of Water" "Erosion" "Ways to Represent Land and Water" **Videos** *All about Soil* *All about Landforms*	**Embedded Assessment** Performance assessment Science notebook entries **Benchmark Assessment** *Investigation 4 I-Check* **NGSS Performance Expectations** 2-ESS1-1 2-ESS2-1; 2-ESS2-2; 2-ESS2-3 K–2 ETS1-1; K–2 ETS1-2; K–2 ETS1-3

2 Full Option Science System | *Pebbles, Sand, and Silt Module—FOSS Next Generation* 3

Module matrix

Framework and the NGSS provides a complete overview of NGSS connections, learning progressions, and background to support the conceptual framework for the module.

The **Materials** chapter is a must-read resource that helps you get your student equipment ready for first-time use and shares helpful tips for getting your classroom ready for FOSS.

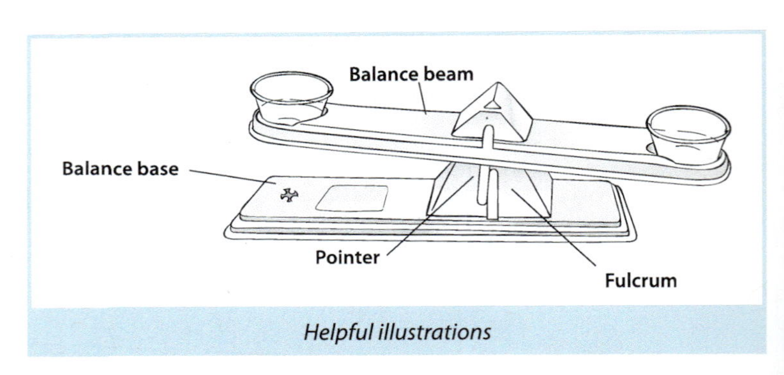

Helpful illustrations

The **Technology** chapter provides an overview for each digital resource in the module and gets you up and running on FOSSweb.com, complete with technical support.

Each **Investigation** includes an At-a-Glance overview, science background content with NGSS connections, and in-depth guidance for preparing and facilitating instruction.

INVESTIGATION 1 — *First Rocks* | *At a Glance*

	Investigation Summary	Time	Focus Question for Phenomenon, Practices	Content Related to DCIs	Writing/Reading	Assessment
PART 1	**Three Rocks** Students investigate and sort a set of six rocks. They gather information about the rocks by observing and comparing, then rub them together to simulate weathering.	**Active Inv.** 1 Session*	What happens when rocks rub together? **Practices** Planning and carrying out investigations Analyzing and interpreting data Constructing explanations	• Rocks can be described by their properties. • Smaller rocks (sand) result from the breaking (weathering) of larger rocks.	**Science Notebook Entry** *Rubbing Rocks*	**Embedded Assessment** Science notebook entry
PART 2	**Washing Three Rocks** Students wash their rocks to see how they change when they are wet, and to see what happens to the wash water. Students are introduced to the names of these volcanic rocks (tuff, scoria, basalt) and view a video on volcanoes to find out how they formed.	**Active Inv.** 1 Session	What happens when rocks are placed in water? **Practices** Planning and carrying out investigations Analyzing and interpreting data Constructing explanations Obtaining, evaluating, and communicating information	• Rocks can be described by their properties. • When rocks are washed in water, the colors or sparkling qualities are enhanced. • Some rocks (such as tuff, scoria, and basalt) are formed from lava and other materials produced by erupting volcanoes. • Volcanoes are mountains built by melted rock that flow out of weak areas in Earth's crust.	**Science Notebook Entry** *Rocks in Water* **Video** *All about Volcanoes*	**Embedded Assessment** Science notebook entry
PART 3	**First Sorting** Students are introduced to river rocks, describe their properties, and compare and sort them into groups based on one property at a time.	**Active Inv.** 1 Session	How are river rocks the same? **Practices** Planning and carrying out investigations Analyzing and interpreting data Constructing explanations Engaging in argument from evidence	• Rocks can be sorted by their properties. • When rocks are washed in water, the colors or sparkling qualities are enhanced.	**Science Notebook Entry** *Rock Sorting*	**Embedded Assessment** Performance assessment
PART 4	**Start a Rock Collection** Students take a field trip to collect and observe schoolyard rocks. They describe the properties of the various rocks.	**Active Inv.** 2 Sessions **Reading** 1 Session	What are the properties of schoolyard rocks? **Practices** Asking questions Analyzing and interpreting data Obtaining, evaluating, and communicating information	• Rocks are all around us. • Rocks are the solid material of Earth.	**Science Notebook Entry** *Rock Record* **Science Resources Book** "Exploring Rocks"	**Embedded Assessment** Science notebook entry
PART 5	**Sorting Activities** Students use sorting mats to compare and sort the river rocks.	**Active Inv.** 1 Session **Reading** 1 Session **Assessment** 1 Session	How many ways can rocks be sorted? **Practices** Analyzing and interpreting data Obtaining, evaluating, and communicating information	• Rocks can be described by their properties. • Rocks are composed of minerals.	**Science Resources Book** "Colorful Rocks" **Online Activities** "Rock Sorting" "Property Chain"	**Benchmark Assessment** *Investigation 1 I-Check* **NGSS Performance Expectations addressed in this investigation** 2-ESS1-1 2-PS1-1

* A class session is 45–50 minutes.

70 Full Option Science System | *Pebbles, Sand, and Silt Module—FOSS Next Generation* 71

The At-a-Glance chart includes:

- Summaries and paci
for investigation scheduling
- Focus questions for investigative phenomena
- Connections to disciplinary core ide
- Reading, writing, an
technology integrat
opportunities
- Embedded and benchmark assessments

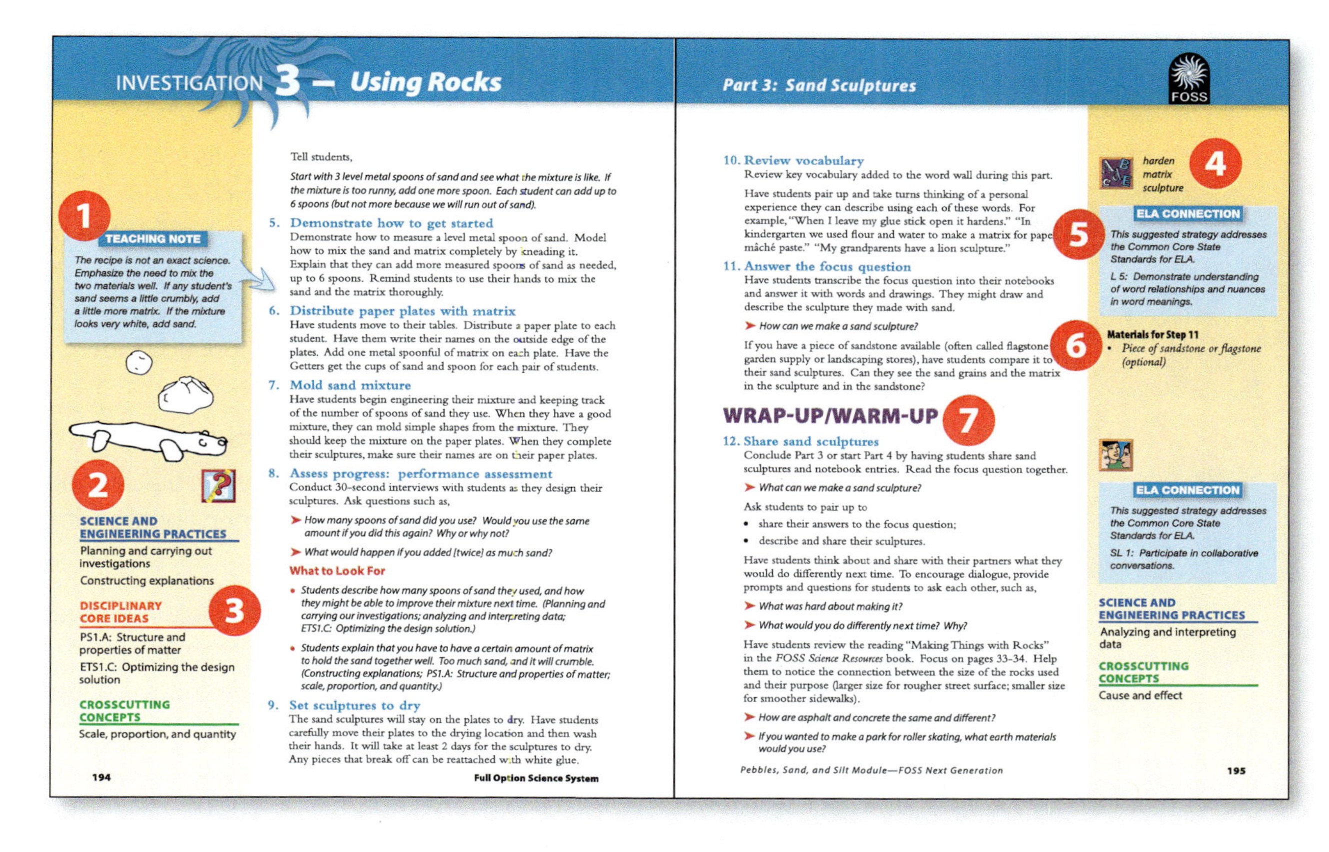

INVESTIGATION 3 — Using Rocks

Part 3: Sand Sculptures

1

TEACHING NOTE

The recipe is not an exact science. Emphasize the need to mix the two materials well. If any student's sand seems a little crumbly, add a little more matrix. If the mixture looks very white, add sand.

2

SCIENCE AND ENGINEERING PRACTICES

Planning and carrying out investigations

Constructing explanations

3

DISCIPLINARY CORE IDEAS

PS1.A: Structure and properties of matter

ETS1.C: Optimizing the design solution

CROSSCUTTING CONCEPTS

Scale, proportion, and quantity

Tell students,

Start with 3 level metal spoons of sand and see what the mixture is like. If the mixture is too runny, add one more spoon. Each student can add up to 6 spoons (but not more because we will run out of sand).

5. **Demonstrate how to get started**
Demonstrate how to measure a level metal spoon of sand. Model how to mix the sand and matrix completely by kneading it. Explain that they can add more measured spoons of sand as needed, up to 6 spoons. Remind students to use their hands to mix the sand and the matrix thoroughly.

6. **Distribute paper plates with matrix**
Have students move to their tables. Distribute a paper plate to each student. Have them write their names on the outside edge of the plates. Add one metal spoonful of matrix on each plate. Have the Getters get the cups of sand and spoon for each pair of students.

7. **Mold sand mixture**
Have students begin engineering their mixture and keeping track of the number of spoons of sand they use. When they have a good mixture, they can mold simple shapes from the mixture. They should keep the mixture on the paper plates. When they complete their sculptures, make sure their names are on their paper plates.

8. **Assess progress: performance assessment**
Conduct 30-second interviews with students as they design their sculptures. Ask questions such as,

➤ *How many spoons of sand did you use? Would you use the same amount if you did this again? Why or why not?*

➤ *What would happen if you added [twice] as much sand?*

What to Look For

- *Students describe how many spoons of sand they used, and how they might be able to improve their mixture next time. (Planning and carrying our investigations; analyzing and interpreting data; ETS1.C: Optimizing the design solution.)*
- *Students explain that you have to have a certain amount of matrix to hold the sand together well. Too much sand, and it will crumble. (Constructing explanations; PS1.A: Structure and properties of matter; scale, proportion, and quantity.)*

9. **Set sculptures to dry**
The sand sculptures will stay on the plates to dry. Have students carefully move their plates to the drying location and then wash their hands. It will take at least 2 days for the sculptures to dry. Any pieces that break off can be reattached with white glue.

194 Full Option Science System

10. **Review vocabulary**
Review key vocabulary added to the word wall during this part.
Have students pair up and take turns thinking of a personal experience they can describe using each of these words. For example, "When I leave my glue stick open it hardens." "In kindergarten we used flour and water to make a matrix for papier-mâché paste." "My grandparents have a lion sculpture."

11. **Answer the focus question**
Have students transcribe the focus question into their notebooks and answer it with words and drawings. They might draw and describe the sculpture they made with sand.

➤ *How can we make a sand sculpture?*

If you have a piece of sandstone available (often called flagstone at garden supply or landscaping stores), have students compare it to their sand sculptures. Can they see the sand grains and the matrix in the sculpture and in the sandstone?

WRAP-UP/WARM-UP 7

12. **Share sand sculptures**
Conclude Part 3 or start Part 4 by having students share sand sculptures and notebook entries. Read the focus question together.

➤ *What can we make a sand sculpture?*

Ask students to pair up to

- share their answers to the focus question;
- describe and share their sculptures.

Have students think about and share with their partners what they would do differently next time. To encourage dialogue, provide prompts and questions for students to ask each other, such as,

➤ *What was hard about making it?*

➤ *What would you do differently next time? Why?*

Have students review the reading "Making Things with Rocks" in the *FOSS Science Resources* book. Focus on pages 33–34. Help them to notice the connection between the size of the rocks used and their purpose (larger size for rougher street surface; smaller size for smoother sidewalks).

➤ *How are asphalt and concrete the same and different?*

➤ *If you wanted to make a park for roller skating, what earth materials would you use?*

4

harden
matrix
sculpture

5

ELA CONNECTION

This suggested strategy addresses the Common Core State Standards for ELA.

L 5: Demonstrate understanding of word relationships and nuances in word meanings.

6

Materials for Step 11

- *Piece of sandstone or flagstone (optional)*

ELA CONNECTION

This suggested strategy addresses the Common Core State Standards for ELA.

SL 1: Participate in collaborative conversations.

SCIENCE AND ENGINEERING PRACTICES

Analyzing and interpreting data

CROSSCUTTING CONCEPTS

Cause and effect

Pebbles, Sand, and Silt Module—FOSS Next Generation 195

FOSS investigations provide the right support, when you need it with point-of-use guidance.

1. Teaching notes from real classrooms
2. Key three-dimensional highlights
3. Embedded assessment "What to Look For" in grades 1–2
4. Vocabulary review
5. Strategies to support English Language Arts
6. Materials used in the current steps
7. Guiding questions to help students make connections

The **Assessment** chapter gives you an in-depth look at the research-based components of the FOSS Assessment System, guidance on assessing for the NGSS, and generalized next-step strategies to use in your classroom. Find duplication masters, assessment charts, coding guides, and specific next-step strategies on FOSSweb.com.

Getting Started with Your Teacher Toolkit

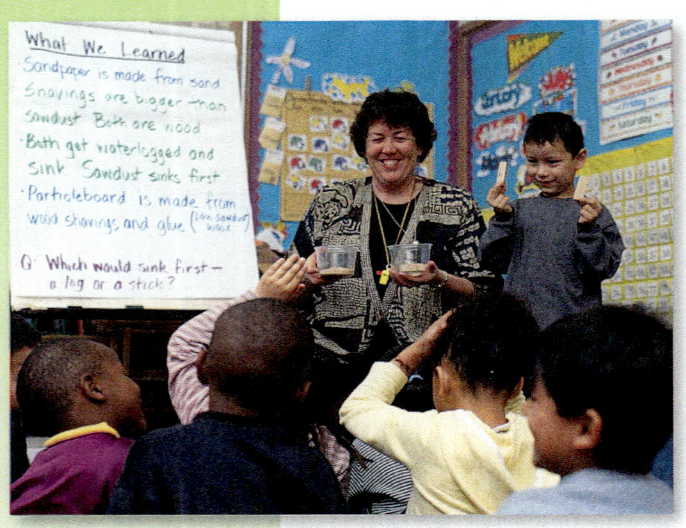

2. Teacher Resources

Your *Investigations Guide* tells you how to facilitate each investigation of a module. The ***Teacher Resources*** provides guidance on how to do it at your grade level across three modules throughout the year with effective practices and strategies derived from extensive field-testing.

A grade-level **Planning Guide** provides an overview to your three modules and an introduction to three-dimensional teaching and learning.

The **Science Notebooks** chapter provides age-appropriate methods to support students in developing productive science notebooks. Access powerful research-based next-step strategies to maximize the effectiveness of the notebook as a formative assessment tool.

Science-Centered Language Development is a collection of standards-aligned strategies to support and enhance literacy development in the context of science—reading, writing, speaking, listening, and vocabulary development.

In **Taking FOSS Outdoors**, find guidance for managing the space, time, and materials needed to provide authentic, real-world learning experiences in students' local communities.

Teacher Resources also includes:

- Grade-level connections to Common Core ELA and Math standards
- Module-specific notebook, teacher, and assessment blackline masters.

Check FOSSweb for the latest updates to chapters in *Teacher Resources*.

3. *FOSS Science Resources* Student Book

The Teacher Toolkit includes one copy of the student book. Reading is an integral part of science learning. Reading informational text critically and effectively is an important component of today's ELA standards. Once students have engaged with phenomena firsthand, they go more in-depth with articles in *FOSS Science Resources*.

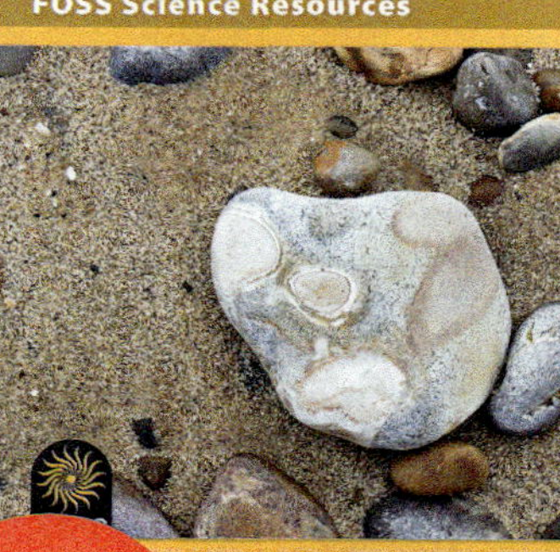

Module includes Big Book

Articles from FOSS *Science Resources* complement and enhance the active investigations, giving students opportunities to:

- Ask and answer questions
- Use evidence to support their ideas
- Use text to acquire information
- Draw information from multiple sources
- Interpret illustrations to build understanding

Erosion

What happened to this road? People once drove on this road. During a big storm, waves crashed against the shore. They washed away the soil under the road. Parts of the road were destroyed.

Waves are moving water that cause **erosion** on a coast. Ocean waves often erode the shore during storms. Coastal erosion can damage roads and buildings. Waves can also wash away all the sand on a beach.

68 69

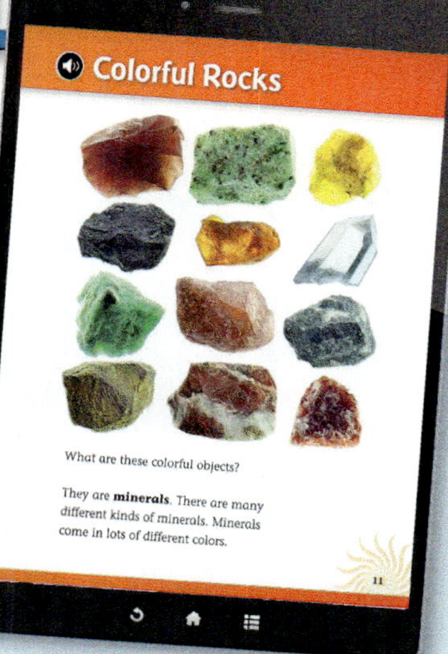

Colorful Rocks
What are these colorful objects?
They are minerals. There are many different kinds of minerals. Minerals come in lots of different colors.
11

Interactive eBooks

FOSS Science Resources is available as a convenient, platform-neutral interactive student eBook with integrated audio, highlighted text, and links to videos and online activities. Student access to eBooks is available as an additional purchase.

Getting Started with Technology

FOSSweb.com

Easy access to program support resources

FOSSweb.com is your home for accessing the complete portfolio of digital resources in the FOSS program. Easily manage each of your modules, create class pages, and keep helpful references at your fingertips.

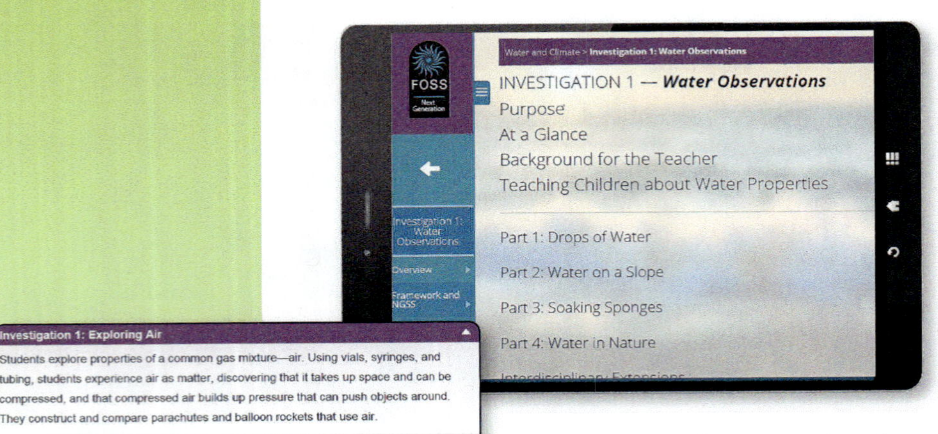

eInvestigations Guide

This easy-to-use interactive version of the *Investigations Guide* is mobile-friendly and offers simplified navigation, collapsible sections, and the ability to add customized notes.

Resources by Investigation

Easily access the duplication masters, online activities, and streaming videos needed for the current investigation part.

Teacher Preparation Videos

Videos provide helpful equipment setup instructions, safety information, and a summary of what students will do and learn throughout a part.

Interactive Whiteboard Lessons

Developed for SMART™ or Promethean boards, these resources help you facilitate each part of every investigation and give the class a visual reference.

Online Activities for Differentiating Instruction

FOSSweb digital resources provide engaging, interactive virtual investigations and tutorials that offer additional content and skill support for students. These experiences also help students who were absent catch up with class.

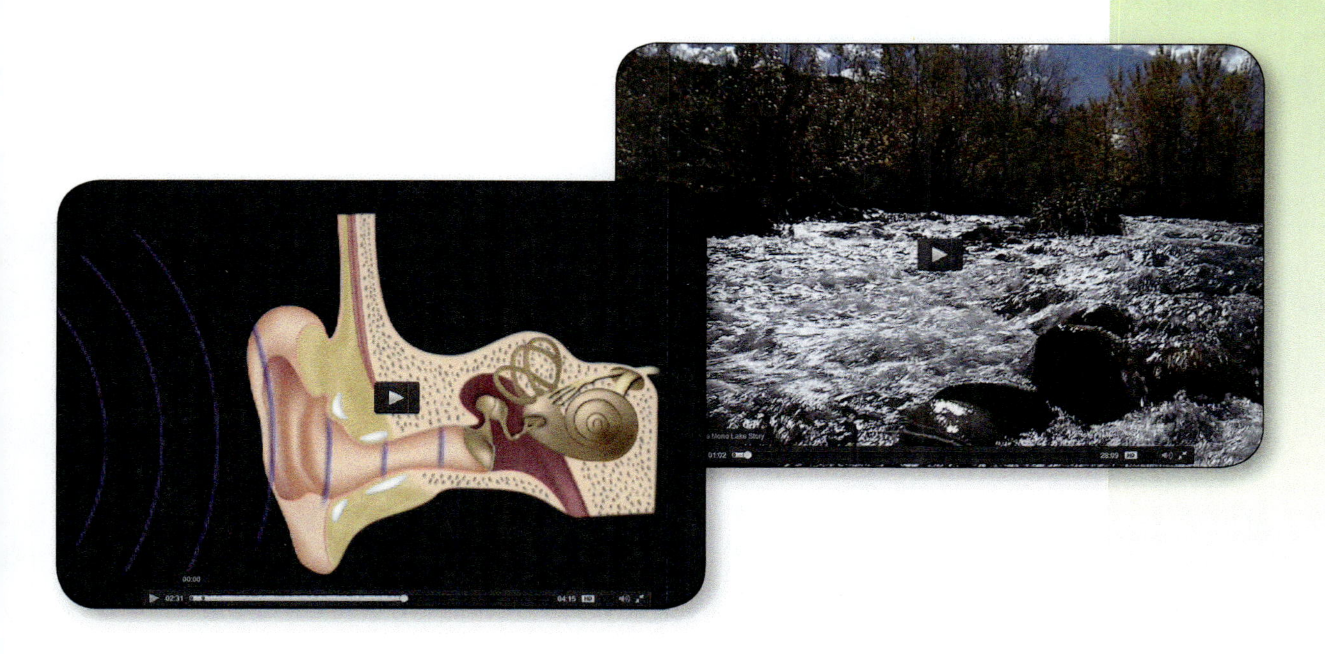

Streaming Videos

Videos are available on FOSSweb to support many investigations and often take students "on location" around the world or showcase experiments that would be too messy, expensive, or dangerous for the classroom.

Three-Dimensional Active Learning

The FOSS program has always placed student learning of science *practices* on equal footing with science *concepts and principles* and the NGSS and *Framework for K–12 Science Education* have provided a new language with which to articulate this. In each **FOSS Next Generation** investigation, students are engaged in the three dimensions of the NGSS to develop increasingly complex knowledge and understanding.

Science & Engineering
PRACTICES
Disciplinary
CORE IDEAS
CROSSCUTTING
Concepts
Science and engineering practices are the cognitive tools scientists and engineers use to answer questions and design solutions. FOSS students use these tools to gather evidence used to explain real-world phenomena.
Grade-level appropriate **disciplinary core ideas** are the concepts and established ideas of science. FOSS students develop these building blocks throughout investigations to make sense of phenomena.
Crosscutting concepts help students to connect the varied concepts and disciplines of science. FOSS students apply these concepts to different situations in order to make connections and develop comprehensive understanding.

FOSS Forward Thinking

The Full Option Science System has evolved into a fully realized 21st century science program with authentic connection to the Next Generation Science Standards (NGSS).

The FOSS Vision

When the Full Option Science System (FOSS) began, the founders envisioned a science curriculum that was enjoyable, logical, and intuitive for teachers, and stimulating, provocative, and informative for students. Achieving this vision was informed by research in cognitive science, learning theory, and critical study of effective practice. The modular design of the FOSS product allowed users to select topics that aligned with district or state learning objectives, or simply resonated with their perception of comprehensive and reasonable science instruction. The original design of the FOSS Program was comprehensive in terms of coverage. FOSS was designed to provide real and meaningful student experience with important scientific ideas and to nurture developmentally appropriate knowledge of the objects, organisms, systems, and principles governing, the natural world.

The FOSS Next Generation Program

But the developers never envisioned FOSS to be a static curriculum, and now the Full Option Science System has evolved into a fully realized 21st century science program with authentic connection to the *Next Generation Science Standards (NGSS)*. The FOSS science curriculum is a comprehensive science program, featuring instructional guidance, student equipment, student reading materials, digital resources, and an embedded assessment system. The FOSS philosophy has always taken very seriously the teaching of good, comprehensive, accurate, science content using the methods of inquiry to advance that science knowledge. But the *Framework for K–12 Science Education*, on which the NGSS are based has allowed us to articulate our mission in a more coherent manner, using the vocabulary established by the authors of the *Framework*. The FOSS instructional design now strives to

a. communicate the disciplinary core ideas (content) of science, while

b. guiding and encouraging students to engage in or exercise the science and engineering practices (inquiry methods) to develop knowledge of the disciplinary core ideas, and

c. help students apprehend the crosscutting concepts (themes that unite core ideas, overarching concepts) that connect the learning experiences within a discipline and bridge meaningfully across disciplines as students gain more and more knowledge of the natural world.

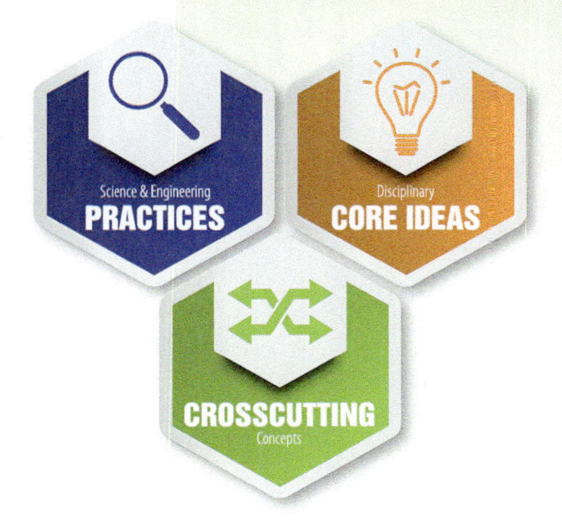

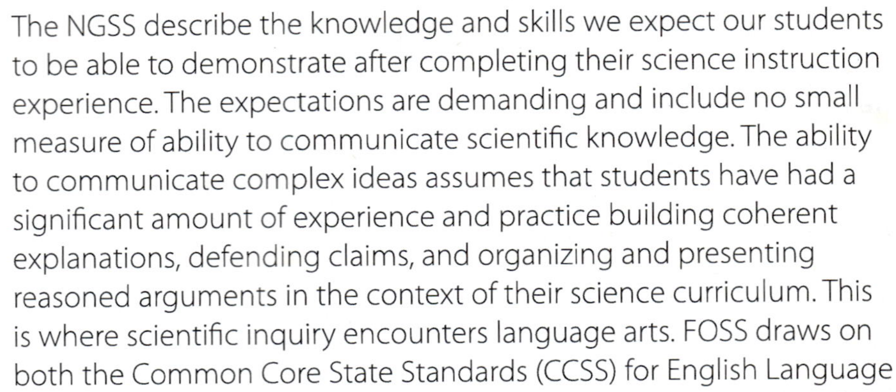

The NGSS describe the knowledge and skills we expect our students to be able to demonstrate after completing their science instruction experience. The expectations are demanding and include no small measure of ability to communicate scientific knowledge. The ability to communicate complex ideas assumes that students have had a significant amount of experience and practice building coherent explanations, defending claims, and organizing and presenting reasoned arguments in the context of their science curriculum. This is where scientific inquiry encounters language arts. FOSS draws on both the Common Core State Standards (CCSS) for English Language Arts and research data regarding the productive use of student science notebooks. FOSS developers realize that the most effective science program must seamlessly integrate science instruction goals and language arts skills. Science is one of the most engaging and productive arenas for introducing and exercising language arts skills: vocabulary, nonfiction (informational) reading, cause-and-effect relationships, on and on.

FOSS is strongly grounded in the realities of the classroom and the interests and experiences of the learners. The content in FOSS is teachable and learnable over multiple grade levels as students increase in their abilities to reason about and integrate complex ideas within and between disciplines.

FOSS is crafted with a structured, yet flexible, teaching philosophy that embraces the much-heralded 21st century skills; collaborative teamwork, critical thinking, and problem solving. The FOSS curriculum design promotes a classroom culture that allows both teachers and students to assume prominent roles in the management of the learning experience.

FOSS is built on the assumptions that understanding of core scientific knowledge and how science functions is essential for citizenship, that all teachers can teach science, and that all students can learn science. Formative assessment in FOSS creates a community of reflective practice. Teachers and students make up the community and establish norms of mutual support, trust, respect, and collaboration. The goal of the community is that everyone will demonstrate progress and will learn and grow.

Overview

INTRODUCTION

Students engage with the anchor phenomenon of earth materials that cover the planet's surface. They observe the properties of rocks of various sizes and study the components of soil, study the results of weathering and erosion, locate natural sources of water, and determine how to represent the shapes and kinds of land and bodies of water on Earth. The guiding questions are what are the properties of earth materials? and how do they interact and change?

Students use simple tools to observe, describe, analyze, and sort solid earth materials and learn how the properties of the materials are suited to different purposes. The investigations compliment the students' experiences in the **Solids and Liquids Module** with a focus on earth materials and the influence of engineering and science on society and the natural world. Students explore how wind and water change the shape of the land and compare ways to slow the process of erosion. Students learn about the important role that earth materials have as natural resources.

Throughout the **Pebbles, Sand, and Silt Module**, students engage in science and engineering practices to collect and interpret data to answer science questions, develop models to communicate interactions and processes, and define problems in order to compare solutions. Students gain experiences that will contribute to understanding of crosscutting concepts of patterns; cause and effect; scale, proportion, and quantity; energy and matter; and stability and change.

Contents

The NGSS Performance Expectations bundled in this module include:

Earth and Space Sciences
2-ESS1-1
2-ESS2-1
2-ESS2-2
2-ESS2-3

Physical Sciences
2-PS1-1
2-PS1-2

Engineering, Technology, and Applications of Science
K–2 ETS1-1
K–2 ETS1-2
K–2 ETS1-3

NOTE
The three modules for grade 2 in FOSS Next Generation are

Solids and Liquids

Pebbles, Sand, and Silt

Insects and Plants

	Investigation Summary	Guiding and Focus Questions for Phenomena
Inv. 1: First Rocks	Students are introduced to the phenomenon that rocks are not all the same. They investigate several kinds of volcanic rocks and begin to understand the properties of rocks. Students observe rocks (using hand lenses), rub rocks, wash rocks, sort rocks, and describe rocks. After rubbing two samples together, students find that rock is hard but also susceptible to weathering. Students also begin to organize a class rock collection.	***What are properties of rocks and how do they change?*** **What happens when rocks rub together?** **What happens when rocks are placed in water?** **How are river rocks the same?** **What are the properties of schoolyard rocks?** **How many ways can rocks be sorted?**
Inv. 2: River Rocks	Students investigate a mixture of different-sized river rocks as a phenomenon. They separate the rocks using a series of three screens to identify five sizes of rocks: large pebbles, small pebbles, large gravel, small gravel, and sand. They add water to a vial of sand to discover silt and clay. Students learn how sand is formed and compare slow landform changes of weathering and erosion to rapid landform changes due to volcanic eruptions.	***How are small pieces of rock made and moved to change landforms?*** **How can rocks be separated by size?** **How else can rocks be sorted by size?** **Is there an earth material smaller than sand?** **What earth material is smaller than silt?** **How does water and wind change landforms?**
Inv. 3: Using Rocks	Students learn how people use earth materials to construct objects. They make rubbings from sandpaper, sculptures from sand, decorative jewelry from clay, and bricks from clay soil. They go on a schoolyard field trip to look for places where earth materials occur naturally and where people have incorporated earth materials into building materials. Students discover that rock as a resource is a natural phenomenon occurring in predictable locations all over Earth's surface.	***How are different sizes of rock used as resources to make useful objects?*** **How do people use earth materials?** **What does sand do for sandpaper?** **How can we make a sand sculpture?** **What makes clay useful in making objects like beads?** **How are bricks made?**
Inv. 4: Soil and Water	Students first investigate a common phenomenon on the surface of Earth—soil. They put together and take apart soils. They are introduced to humus as an ingredient in soil. Homemade and local soils are compared, using techniques introduced in Investigation 2, including water. Students read about sources of natural water, sort images of water sources, both fresh and salt, and discuss where water is found in their community. Students compare different solutions presented in readings to slow the effects of wind and water erosion. They learn about different ways to represent landforms and bodies of water.	***How can we apply what we know about the ways that land and water interact?*** **What is soil?** **How do soils differ?** **Where is water found in our community?** **How can soil erosion be reduced?**

Content Related to Disciplinary Core Ideas	Reading/Technology	Assessment
• Rocks can be described by their properties. • Smaller rocks (sand) result from the breaking (weathering) of larger rocks. • Rocks are the solid material of Earth. • Rocks are composed of minerals. • Volcanoes are mountains built up by melted rocks that flow out of weak areas in Earth's crust.	**Science Resources Book** "Exploring Rocks" "Colorful Rocks" **Video** *All about Volcanoes* **Online Activities** "Rock Sorting" "Property Chain"	**Embedded Assessment** Science notebook entries Performance assessment **Benchmark Assessment** *Investigation 1 I-Check* **NGSS Performance Expectations** 2-ESS1-1 2-PS1-1
• Rocks are earth materials. • Rocks can be described by the property of size. • Rock sizes include clay, silt, sand, gravel, pebbles, cobbles, and boulders. • Weathering, caused by wind or water, causes larger rocks to break into small rocks. • Some Earth events happen rapidly; others occur slowly over a very long period of time.	**Science Resources Book** "The Story of Sand" "Rocks Move" "Landforms" **Video** *All about Land Formations*	**Embedded Assessment** Performance assessment Science notebook entries **Benchmark Assessment** *Investigation 2 I-Check* **NGSS Performance Expectations** 2-ESS1-1 2-ESS2-1; 2-ESS2-2; 2-ESS2-3 2-PS1-1
• Earth materials are natural resources. • The properties of different earth materials make each suitable for specific uses. • Different sizes of sand are used on sandpaper to change the surface of wood from rough to smooth. • Earth materials are commonly used in the construction of buildings and streets.	**Science Resources Book** "Making Things with Rocks" "What Are Natural Resources?" **Online Activity** "Find Earth Materials"	**Embedded Assessment** Science notebook entries Performance assessment **Benchmark Assessment** *Investigation 3 I-Check* **NGSS Performance Expectations** 2-PS1-1 ; 2-PS1-2 K–2 ETS1-1; K–2 ETS1-2; K–2 ETS1-3
• Earth materials are natural resources. • Soils can be described by their properties (color, texture, ability to support plant growth). • Soil is made partly from weathered rock and partly from organic material. Soils vary by location. • Natural sources of water include streams, rivers, ponds, lakes, marshes, and the ocean. Sources of water can be fresh or salt water. • Water can be a solid, liquid, or gas. • Wind and water can change the shape of land. • The shapes and kinds of land and water can be represented by various models.	**Science Resources Book** "What Is in Soil?" "Testing Soil" "Where Is Water Found?" "States of Water" "Erosion" "Ways to Represent Land and Water" **Videos** *All about Soil* *All about Landforms*	**Embedded Assessment** Performance assessment Science notebook entries **Benchmark Assessment** *Investigation 4 I-Check* **NGSS Performance Expectations** 2-ESS1-1 2-ESS2-1; 2-ESS2-2; 2-ESS2-3 K–2 ETS1-1; K–2 ETS1-2; K–2 ETS1-3

FOSS COMPONENTS

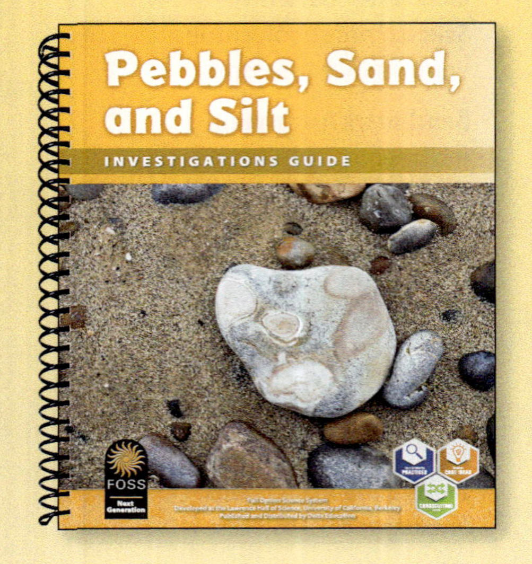

Teacher Toolkit for Each Module

The FOSS Next Generation Program has three modules for grade 2—Solids and Liquids; Pebbles, Sand, and Silt; and Insects and Plants.

Each module comes with a *Teacher Toolkit* for that module. The *Teacher Toolkit* is the most important part of the FOSS Program. It is here that all the wisdom and experience contributed by hundreds of educators has been assembled. Everything we know about the content of the module, how to teach the subject, and the resources that will assist the effort are presented here. Each toolkit has three parts.

Investigations Guide. This spiral-bound document contains these chapters.

- Overview
- Framework and NGSS
- Materials
- Technology
- Investigations (four in this module)
- Assessment

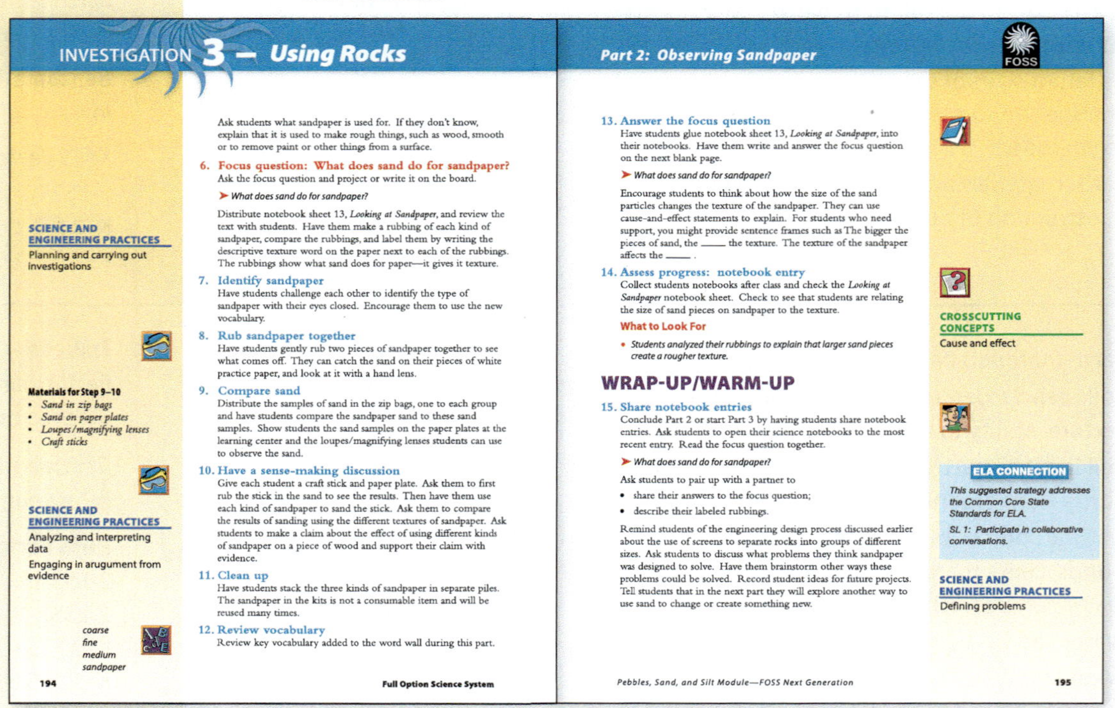

INVESTIGATION 3 — Using Rocks

Ask students what sandpaper is used for. If they don't know, explain that it is used to make rough things, such as wood, smooth or to remove paint or other things from a surface.

6. **Focus question: What does sand do for sandpaper?**
Ask the focus question and project or write it on the board.

➤ *What does sand do for sandpaper?*

Distribute notebook sheet 13, *Looking at Sandpaper*, and review the text with students. Have them make a rubbing of each kind of sandpaper, compare the rubbings, and label them by writing the descriptive texture word on the paper next to each of the rubbings. The rubbings show what sand does for paper—it gives it texture.

SCIENCE AND ENGINEERING PRACTICES
Planning and carrying out investigations

7. **Identify sandpaper**
Have students challenge each other to identify the type of sandpaper with their eyes closed. Encourage them to use the new vocabulary.

8. **Rub sandpaper together**
Have students gently rub two pieces of sandpaper together to see what comes off. They can catch the sand on their pieces of white practice paper, and look at it with a hand lens.

Materials for Step 9–10
- *Sand in zip bags*
- *Sand on paper plates*
- *Loupes/magnifying lenses*
- *Craft sticks*

9. **Compare sand**
Distribute the samples of sand in the zip bags, one to each group and have students compare the sandpaper sand to these sand samples. Show students the sand samples on the paper plates at the learning center and the loupes/magnifying lenses students can use to observe the sand.

10. **Have a sense-making discussion**
Give each student a craft stick and paper plate. Ask them to first rub the stick in the sand to see the results. Then have them use each kind of sandpaper to sand the stick. Ask them to compare the results of sanding using the different textures of sandpaper. Ask students to make a claim about the effect of using different kinds of sandpaper on a piece of wood and support their claim with evidence.

SCIENCE AND ENGINEERING PRACTICES
Analyzing and interpreting data
Engaging in arugument from evidence

11. **Clean up**
Have students stack the three kinds of sandpaper in separate piles. The sandpaper in the kits is not a consumable item and will be reused many times.

12. **Review vocabulary**
Review key vocabulary added to the word wall during this part.

coarse
fine
medium
sandpaper

194 Full Option Science System

Part 2: Observing Sandpaper

FOSS

13. **Answer the focus question**
Have students glue notebook sheet 13, *Looking at Sandpaper*, into their notebooks. Have them write and answer the focus question on the next blank page.

➤ *What does sand do for sandpaper?*

Encourage students to think about how the size of the sand particles changes the texture of the sandpaper. They can use cause-and-effect statements to explain. For students who need support, you might provide sentence frames such as The bigger the pieces of sand, the _____ the texture. The texture of the sandpaper affects the _____ .

14. **Assess progress: notebook entry**
Collect students notebooks after class and check the *Looking at Sandpaper* notebook sheet. Check to see that students are relating the size of sand pieces on sandpaper to the texture.

What to Look For

- *Students analyzed their rubbings to explain that larger sand pieces create a rougher texture.*

CROSSCUTTING CONCEPTS
Cause and effect

WRAP-UP/WARM-UP

15. **Share notebook entries**
Conclude Part 2 or start Part 3 by having students share notebook entries. Ask students to open their science notebooks to the most recent entry. Read the focus question together.

➤ *What does sand do for sandpaper?*

Ask students to pair up with a partner to
- share their answers to the focus question;
- describe their labeled rubbings.

Remind students of the engineering design process discussed earlier about the use of screens to separate rocks into groups of different sizes. Ask students to discuss what problems they think sandpaper was designed to solve. Have them brainstorm other ways these problems could be solved. Record student ideas for future projects. Tell students that in the next part they will explore another way to use sand to change or create something new.

ELA CONNECTION
This suggested strategy addresses the Common Core State Standards for ELA.
SL 1: Participate in collaborative conversations.

SCIENCE AND ENGINEERING PRACTICES
Defining problems

Pebbles, Sand, and Silt Module—FOSS Next Generation 195

***FOSS Science Resources* book.** One copy of the student book of readings is included in the *Teacher Toolkit*.

Teacher Resources. These chapters can be downloaded from FOSSweb and are also in the bound *Teacher Resources* book.

- FOSS Program Goals
- Planning Guide—Grade 2
- Science and Engineering Practices—Grade 2
- Crosscutting Concepts—Grade 2
- Sense-Making Discussions for Three-Dimensional Learning—Grade 2
- Access and Equity
- Science Notebooks in Grades K–2
- Science-Centered Language Development
- FOSS and Common Core ELA—Grade 2
- FOSS and Common Core Math—Grade 2
- Taking FOSS Outdoors
- Science Notebook Masters
- Teacher Masters
- Assessment Masters

Pebbles, Sand, and Silt

TEACHER RESOURCES

The Story of Sand

Put these rock names in order by size from the largest to the smallest.

sand cobble gravel boulder pebble

largest ______________

____

____

____

smallest ______________

Tell the story of sand in your own words.

FOSS Next Generation
© The Regents of the University of California
Can be duplicated for classroom or workshop use.

Pebbles, Sand, and Silt Module
Investigation 2: River Rocks
No. 8—Notebook Master

Equipment Kit for Each Module or Grade Level

The FOSS Program provides the materials needed for the investigations in sturdy, front-opening drawer-and-sleeve cabinets. Inside, you will find high-quality materials packaged for a class of 32 students. Consumable materials are supplied for three uses before you need to resupply. Teachers may be asked to supply small quantities of common classroom materials.

Delta Education can assist you with materials management strategies for schools, districts, and regional consortia.

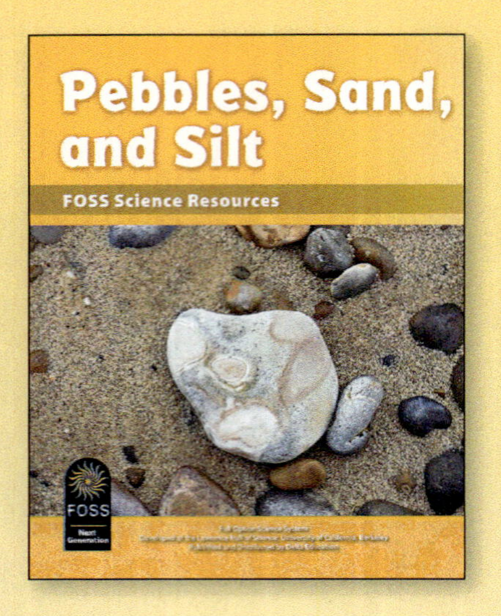

NOTE
FOSS Science Resources: Pebbles, Sand, and Silt is also provided as a big book in the equipment kit.

FOSS Science Resources Books

FOSS Science Resources: Pebbles, Sand, and Silt is a book of original readings developed to accompany this module. The readings are referred to as articles in *Investigations Guide*. Students read the articles in the book as they progress through the module. The articles cover specific concepts, usually after the concepts have been introduced in the active investigation.

The articles in *Science Resources* and the discussion questions provided in *Investigations Guide* help students make connections to the science concepts introduced and explored during the active investigations. Concept development is most effective when students are allowed to experience organisms, objects, and phenomena firsthand before engaging the concepts in text. The text and illustrations help make connections between what students experience concretely and the ideas that explain their observations.

This photograph shows the Ohio River flowing through the city of Columbus, Ohio. This part of the river bank has a park. The park is called Scioto Mile.

This drawing is a different way to show the park.

86

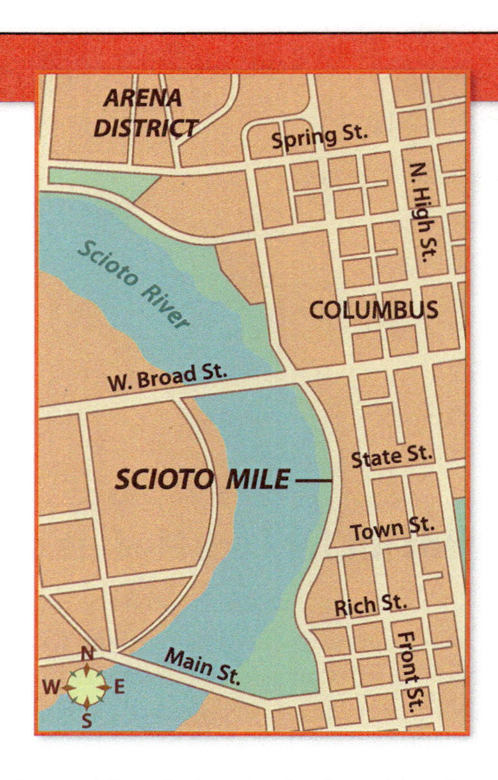

A map of the park shows how big the park is and where it is in Columbus.

87

Technology

The FOSS website opens new horizons for educators, students, and families, in the classroom or at home. Each module has digital resources for students and families—interactive simulations, virtual investigations, and online activities. For teachers, FOSSweb provides online teacher *Investigations Guides*; grade-level planning guides (with connections to ELA and math); materials management strategies; science teaching and professional development tools; contact information for the FOSS Program developers; and technical support. In addition FOSSweb provides digital access to PDF versions of the *Teacher Resources* component of the *Teacher Toolkit*, digital-only instructional resources that supplement the print and kit materials, and access to FOSSmap, the online assessment and reporting system for grades 3–8.

With an educator account, you can customize your homepage, set up easy access to the digital components of the modules you teach, and create class pages for your students with access to online activities.

NOTE
To access all the teacher resources and to set up customized pages for using FOSS, log in to FOSSweb through an educator account. See the Technology chapter in this guide for more specifics.

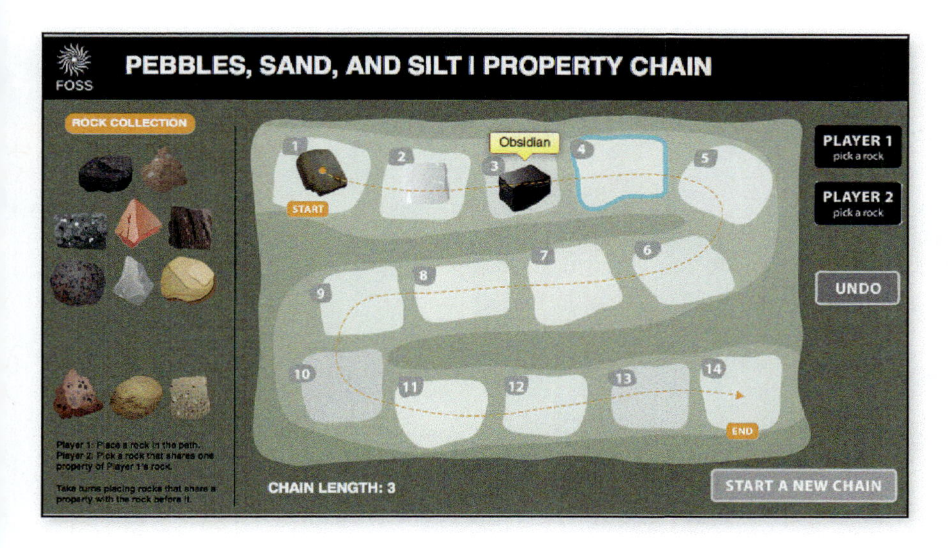

Ongoing Professional Learning

The Lawrence Hall of Science and Delta Education strive to develop long-term partnerships with districts and teachers through thoughtful planning, effective implementation, and ongoing teacher support. FOSS has a strong network of consultants who have rich and experienced backgrounds in diverse educational settings using FOSS.

NOTE
Look for professional development opportunities and online teaching resources on www.FOSSweb.com.

FOSS INSTRUCTIONAL DESIGN

FOSS is designed around active investigation that provides engagement with science concepts and science and engineering practices. Surrounding and supporting those firsthand investigations are a wide range of experiences that help build student understanding of core science concepts and deepen scientific habits of mind.

The Elements of the FOSS Instructional Design

FOSS Instructional Design

Each FOSS investigation follows a similar design to provide multiple exposures to science concepts. The design includes these pedagogies.

- Active investigation in collaborative groups: firsthand experiences with phenomena in the natural and designed worlds
- Recording in science notebooks to answer a focus question dealing with the scientific phenomenon under investigation
- Reading informational text in *FOSS Science Resources* books
- Online activities to acquire data or information or to elaborate and extend the investigation
- Outdoor experiences to collect data from the local environment or to apply knowledge
- Assessment to monitor progress and inform student learning

In practice, these components are seamlessly integrated into a curriculum designed to maximize every student's opportunity to learn.

A **learning cycle** employs an instructional model based on a constructivist perspective that calls on students to be actively involved in their own learning. The model systematically describes both teacher and learner behaviors in a coherent approach to science instruction.

A popular model describes a sequence of five phases of intellectual involvement known as the 5Es: engage, explore, explain, elaborate, and evaluate. The body of foundational knowledge that informs contemporary learning-cycle thinking has been incorporated seamlessly and invisibly into the FOSS curriculum design.

Engagement with real-world **phenomena** is at the heart of FOSS. In every part of every investigation, the investigative phenomenon is referenced implicitly in the focus question that guides instruction and frames the intellectual work. The focus question is a prominent part of each lesson and is called out for the teacher and student. The investigation Background for the Teacher section is organized by focus question—the teacher has the opportunity to read and reflect on the phenomenon in each part in preparing for the lesson. Students record the focus question in their science notebooks, and after exploring the phenomenon thoroughly, explain their thinking in words and drawings.

In science, a phenomenon is a natural occurrence, circumstance, or structure that is perceptible by the senses—an observable reality. Scientific phenomena are not necessarily phenomenal (although they may be)—most of the time they are pretty mundane and well within the everyday experience. What FOSS does to enact an effective engagement with the NGSS is thoughtful selection of scientific phenomena for students to investigate.

NOTE

The anchor phenomena establish the storyline for the module. The investigative phenomena guide each investigation part. Related examples of everyday phenomena are incorporated into the readings, videos, discussions, formative assessments, outdoor experiences, and extensions.

Active Investigation

Active investigation is a master pedagogy. Embedded within active learning are a number of pedagogical elements and practices that keep active investigation vigorous and productive. The enterprise of active investigation includes

- context: sharing prior knowledge, questioning, and planning;
- activity: doing and observing;
- data management: recording, organizing, and processing;
- analysis: discussing and writing explanations.

Context: sharing, questioning, and planning. Active investigation requires focus. The context of an inquiry can be established with a focus question about a phenomenon or challenge from you or, in some cases, from students. (What happens when rocks rub together?) At other times, students are asked to plan a method for investigation. This might start with a teacher demonstration or presentation. Then you challenge students to plan an investigation, such as to find out how rocks can be separated by size. In either case, the field available for thought and interaction is limited. This clarification of context and purpose results in a more productive investigation.

Activity: doing and observing. In the practice of science, scientists put things together and take things apart, observe systems and interactions, and conduct experiments. This is the core of science—active, firsthand experience with objects, organisms, materials, and systems in the natural world. In FOSS, students engage in the same processes. Students often conduct investigations in collaborative groups of four, with each student taking a role to contribute to the effort.

The active investigations in FOSS are cohesive, and build on each other to lead students to a comprehensive understanding of concepts. Through investigations and readings, students gather meaningful data.

Data management: recording, organizing, and processing. Data accrue from observation, both direct (through the senses) and indirect (mediated by instrumentation). Data are the raw material from which scientific knowledge and meaning are synthesized. During and after work with materials, students record data in their science notebooks. Data recording is the first of several kinds of student writing.

Students then organize data so they will be easier to think about. Tables allow efficient comparison. Organizing data in a sequence (time) or series (size) can reveal patterns. Students process some data into graphs, providing visual display of numerical data. They also organize data and process them in the science notebook.

Analysis: discussing and writing explanations. The most important part of an active investigation is extracting its meaning. This constructive process involves logic, discourse, and prior knowledge. Students share their explanations for phenomena, using evidence generated during the investigation to support their ideas. They conclude the active investigation by writing a summary of their learning in their science notebooks, as well as proposing answers to questions raised during the activity.

Science Notebooks

Research and best practice have led FOSS to place more emphasis on the student science notebook. Keeping a notebook helps students organize their observations and data, process their data, and maintain a record of their learning for future reference. The process of writing about their science experiences and communicating their thinking is a powerful learning device for students. The science-notebook entries stand as credible and useful expressions of learning. The artifacts in the notebooks form one of the core exhibitions of the assessment system.

Full-size duplication masters are available on FOSSweb. Student work is entered partly in spaces provided on the notebook sheets and partly on adjacent blank sheets in the composition book. Look to the chapter in *Teacher Resources* called Science Notebooks in Grades K–2 for more details on how to use notebooks with FOSS.

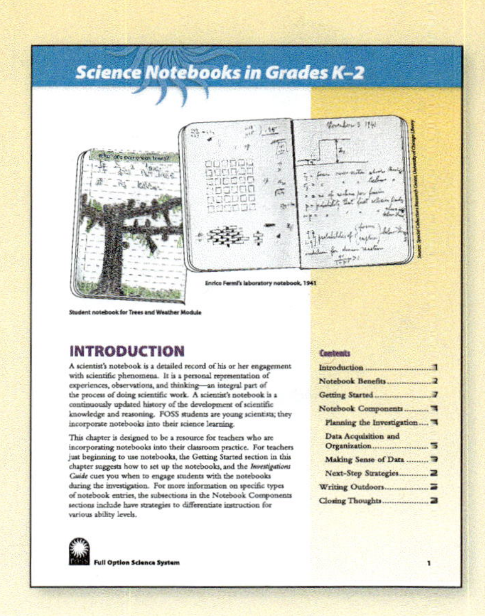

Science Notebooks in Grades K–2

Enrico Fermi's laboratory notebook, 1941

Student notebook for Trees and Weather Module

INTRODUCTION

A scientist's notebook is a detailed record of his or her engagement with scientific phenomena. It is a personal representation of experiences, observations, and thinking—an integral part of the process of doing scientific work. A scientist's notebook is a continuously updated history of the development of scientific knowledge and reasoning. FOSS students are young scientists; they incorporate notebooks into their science learning.

This chapter is designed to be a resource for teachers who are incorporating notebooks into their classroom practice. For teachers just beginning to use notebooks, the Getting Started section in this chapter suggests how to set up the notebooks, and the *Investigations Guide* cues you when to engage students with the notebooks during the investigation. For more information on specific types of notebook entries, the subsections in the Notebook Components sections include have strategies to differentiate instruction for various ability levels.

Contents

Full Option Science System 1

Reading in *FOSS Science Resources*

The *FOSS Science Resources* books, available in print and interactive eBooks, are primarily devoted to expository articles and biographical sketches. FOSS suggests that the reading be completed during language-arts time to connect to the Common Core State Standards for ELA. When language-arts skills and methods are embedded in content material that relates to the authentic experience students have had during the FOSS active learning sessions, students are interested, and they get more meaning from the text material.

Recommended strategies to engage students in reading, writing, speaking, and listening using the articles in the *FOSS Science Resources* books are included in the flow of Guiding the Investigation. In addition, a library of resources is described in the Science-Centered Language Development chapter in *Teacher Resources*.

The FOSS and Common Core ELA—Grade 2 chapter in *Teacher Resources* shows how FOSS provides opportunities to develop and exercise the Common Core ELA practices through science. A detailed table identifies these opportunities in the three FOSS modules for the second grade.

Engaging in Online Activities through FOSSweb

The simulations and online activities on FOSSweb are designed to support students' learning at specific times during instruction. Digital resources include streaming videos that can be viewed by the class or small groups. Resources can be used to review the active investigations and to support students who need more time with the concepts.

The Technology chapter provides details about the online activities for students and the tools and resources for teachers to support and enrich instruction. There are many ways for students to engage with the digital resources—in class as individuals, in small groups, or as a whole class, and at home with family and friends.

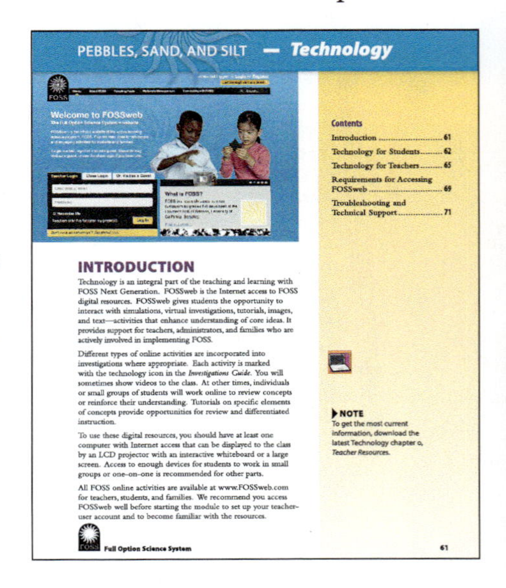

PEBBLES, SAND, AND SILT — *Technology*

Welcome to FOSSweb

Contents

INTRODUCTION

Technology is an integral part of the teaching and learning with FOSS Next Generation. FOSSweb is the Internet access to FOSS digital resources. FOSSweb gives students the opportunity to interact with simulations, virtual investigations, tutorials, images, and text—activities that enhance understanding of core ideas. It provides support for teachers, administrators, and families who are actively involved in implementing FOSS.

Different types of online activities are incorporated into investigations where appropriate. Each activity is marked with the technology icon in the *Investigations Guide*. You will sometimes show videos to the class. At other times, individuals or small groups of students will work online to review concepts or reinforce their understanding. Tutorials on specific elements of concepts provide opportunities for review and differentiated instruction.

To use these digital resources, you should have at least one computer with Internet access that can be displayed to the class by an LCD projector with an interactive whiteboard or a large screen. Access to enough devices for students to work in small groups or one-on-one is recommended for other parts.

All FOSS online activities are available at www.FOSSweb.com for teachers, students, and families. We recommend you access FOSSweb well before starting the module to set up your teacher-user account and to become familiar with the resources.

NOTE
To get the most current information, download the latest Technology chapter o, *Teacher Resources*.

Full Option Science System 61

Assessing Progress

The FOSS assessment system includes both formative and summative assessments. Formative assessment monitors learning during the process of instruction. It measures progress, provides information about learning, and is predominantly diagnostic. Summative assessment looks at the learning after instruction is completed, and it measures achievement.

Formative assessment in FOSS, called **embedded assessment**, is an integral part of instruction, and occurs on a daily basis. You observe action during class in a performance assessment or review notebooks after class. Performance assessments look at students' engagement in science and engineering practices or their recognition of crosscutting concepts, and are indicated with the second assessment icon. Embedded assessment provides continuous monitoring of students' learning and helps you make decisions about whether to review, extend, or move on to the next idea to be covered.

Benchmark assessments are short summative assessments given after each investigation. These **I-Checks** are actually hybrid tools: they provide summative information about students' achievement, and because they occur soon after teaching each investigation, they can be used diagnostically as well. Reviewing specific items on an I-Check with the class provides additional opportunities for students to clarify their thinking.

The embedded assessments are based on authentic work produced by students during the course of participating in the FOSS activities. Students do their science, and you observe the actions and look at their notebook entries. Bullet points in the Guiding the Investigation tell you specifically what students should know and be able to communicate.

If student work is incorrect or incomplete, you know that there has been a breakdown in the learning/communicating process. The assessment system then provides a menu of next-step strategies to resolve the situation. Embedded assessment is assessment *for* learning, not assessment *of* learning.

Assessment *of* learning is the domain of the benchmark assessments. Benchmark assessments for grades 1–2 are delivered after each investigation (I-Checks). These assessments can also be used to monitor and adjust instruction based on student understandings.

The Three Rs of Conservation

A natural resource is something found in nature that people need or use. Trees, soil, and water are some natural resources. Keeping natural resources safe and using them wisely is called conservation. How can you practice conservation? Follow the three Rs: reduce, reuse, and recycle!

REDUCE

REUSE

RECYCLE

NOTE

The kit includes a set of four *Conservation* posters so you can discuss the importance of natural resources with students.

Taking FOSS Outdoors

FOSS throws open the classroom door and proclaims the entire school campus to be the science classroom. The true value of science knowledge is its usefulness in the real world and not just in the classroom. Taking regular excursions into the immediate outdoor environment has many benefits. First of all, it provides opportunities for students to apply things they learned in the classroom to novel situations. When students are able to transfer knowledge of scientific principles to natural systems, they experience a sense of accomplishment.

In addition to transfer and application, students can learn things outdoors that they are not able to learn indoors. The most important object of inquiry outdoors is the outdoors itself. To today's youth, the outdoors is something to pass through as quickly as possible to get to the next human-managed place. For many, engagement with the outdoors and natural systems must be intentional, at least at first. With repeated visits to familiar outdoor learning environments, students may first develop comfort in the outdoors, and then a desire to embrace and understand natural systems.

The last part of most investigations is an outdoor experience. Venturing out will require courage the first time or two you mount an outdoor expedition. It will confuse students as they struggle to find the right behavior that is a compromise between classroom rigor and diligence and the freedom of recreation. With persistence, you will reap rewards. You will be pleased to see students' comportment develop into proper field-study habits, and you might be amazed by the transformation of students with behavior issues in the classroom who become your insightful observers and leaders in the schoolyard environment.

Teaching outdoors is the same as teaching indoors—except for the space. You need to manage the same four core elements of classroom teaching: time, space, materials, and students. Because of the different space, new management procedures are required. Students can get farther away. Materials have to be transported. The space has to be defined and honored. Time has to be budgeted for getting to, moving around in, and returning from the outdoor study site. All these and more issues and solutions are discussed in the Taking FOSS Outdoors chapter in *Teacher Resources*.

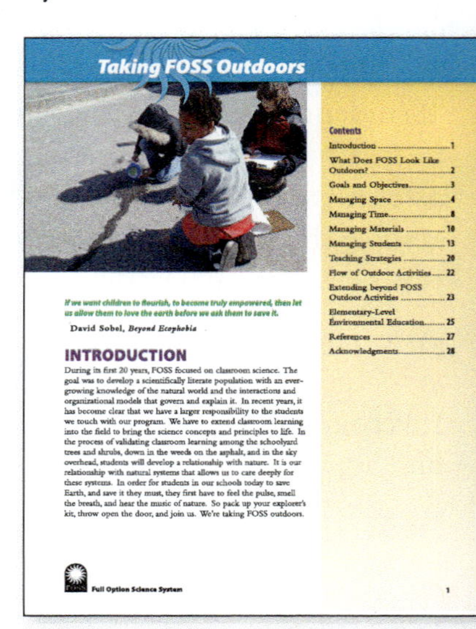

Taking FOSS Outdoors

Contents

If we want children to flourish, to become truly empowered, then let us allow them to love the earth before we ask them to save it.

David Sobel, *Beyond Ecophobia*

INTRODUCTION

During its first 20 years, FOSS focused on classroom science. The goal was to develop a scientifically literate population with an ever-growing knowledge of the natural world and the interactions and organizational models that govern and explain it. In recent years, it has become clear that we have a larger responsibility to the students we touch with our program. We have to extend classroom learning into the field to bring the science concepts and principles to life. In the process of validating classroom learning among the schoolyard trees and shrubs, down in the weeds on the asphalt, and in the sky overhead, students will develop a relationship with nature. It is our relationship with natural systems that allows us to care deeply for these systems. In order for students in our schools today to save Earth, and save it they must, they first have to feel the pulse, smell the breath, and hear the music of nature. So pack up your explorer's kit, throw open the door, and join us. We're taking FOSS outdoors.

Full Option Science System 1

Science-Centered Language Development and Common Core State Standards for ELA

The FOSS active investigations, science notebooks, *FOSS Science Resources* articles, and formative assessments provide rich contexts in which students develop and exercise thinking and communication. These elements are essential for effective instruction in both science and language arts—students experience the natural world in real and authentic ways and use language to inquire, process information, and communicate their thinking about scientific phenomena. FOSS refers to this development of language process and skills within the context of science as science-centered language development.

In the Science-Centered Language Development chapter in *Teacher Resources*, we explore the intersection of science and language and the implications for effective science teaching and language development. Language plays two crucial roles in science learning: (1) it facilitates the communication of conceptual and procedural knowledge, questions, and propositions, and (2) it mediates thinking—a process necessary for understanding. For students, language development is intimately involved in their learning about the natural world. Science provides a real and engaging context for developing literacy and language-arts skills identified in contemporary standards for English language arts.

The most effective integration depends on the type of investigation, the experience of students, the language skills and needs of students, and the language objectives that you deem important at the time. The Science-Centered Language Development chapter is a library of resources and strategies for you to use. The chapter describes how literacy strategies are integrated purposefully into the FOSS investigations, gives suggestions for additional literacy strategies that both enhance students' learning in science and develop or exercise English-language literacy skills, and develops science vocabulary with scaffolding strategies for supporting all learners. We identify effective practices in language-arts instruction that support science learning and examine how learning science content and engaging in science and engineering practices support language development.

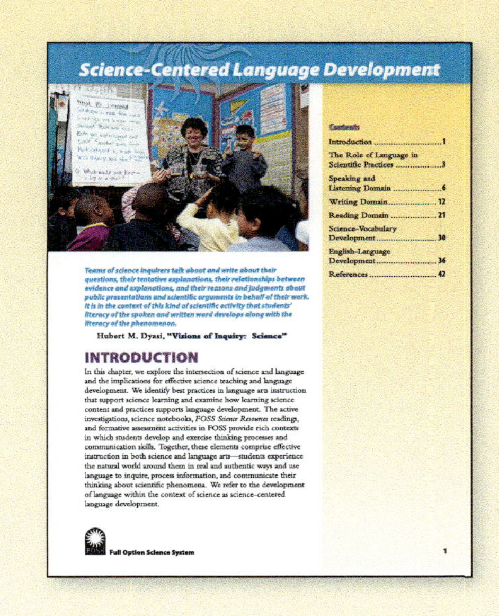

Science-Centered Language Development

Contents

Teams of science inquirers talk about and write about their questions, their tentative explanations, their relationships between evidence and explanations, and their reasons and judgments about public presentations and scientific arguments in behalf of their work. It is in the context of this kind of scientific activity that students' literacy of the spoken and written word develops along with the literacy of the phenomenon.

Hubert M. Dyasi, "Visions of Inquiry: Science"

INTRODUCTION

In this chapter, we explore the intersection of science and language and the implications for effective science teaching and language development. We identify best practices in language arts instruction that support science learning and examine how learning science content and practices supports language development. The active investigations, science notebooks, *FOSS Science Resources* readings, and formative assessment activities in FOSS provide rich contexts in which students develop and exercise thinking processes and communication skills. Together, these elements comprise effective instruction in both science and language arts—students experience the natural world around them in real and authentic ways and use language to inquire, process information, and communicate their thinking about scientific phenomena. We refer to the development of language within the context of science as science-centered language development.

Full Option Science System 1

Specific methods to make connections to the Common Core State Standards for English Language Arts are included in the flow of Guiding the Investigation. These recommended methods are linked to the CCSS ELA through ELA Connection notes. In addition, the FOSS and the Common Core ELA chapter in *Teacher Resources* summarizes all of the connections to each standard at the given grade level.

DIFFERENTIATED INSTRUCTION FOR ACCESS AND EQUITY

Learning from Experience

"Active science by itself provides part of the solution to full inclusion and provides many opportunities at the same time for differentiated instruction."

The roots of FOSS extend back to the mid-1970s and the Science Activities for the Visually Impaired and Science Enrichment for Learners with Physical Handicaps projects (SAVI/SELPH Program). As this special-education science program expanded into fully integrated (mainstreamed) settings in the 1980s, hands-on science proved to be a powerful medium for bringing all students together. The subject matter is universally interesting, and the joy and satisfaction of discovery are shared by everyone. Active science by itself provides part of the solution to full inclusion and provides many opportunities at the same time for differentiated instruction.

Many years later, FOSS began a collaboration with educators and researchers at the Center for Applied Special Technology (CAST), where principles of Universal Design for Learning (UDL) had been developed and applied. FOSS continues to learn from our colleagues about ways to use new media and technologies to improve instruction. Here are the UDL guiding principles.

Principle 1. Provide multiple means of representation. Give learners various ways to acquire information and demonstrate knowledge.

Principle 2. Provide multiple means of action and expression. Offer students alternatives for communicating what they know.

Principle 3. Provide multiple means of engagement. Help learners get interested, be challenged, and stay motivated.

FOSS for All Students

The FOSS Program has been designed to maximize the science learning opportunities for all students, including those who have traditionally not had access to or have not benefited from equitable science experiences—students with special needs, ethnically diverse learners, English learners, students living in poverty, girls, and advanced and gifted learners. FOSS is rooted in a 30-year tradition of multisensory science education and informed by recent research on UDL and culturally and linguistically responsive teaching and learning. Procedures found effective with students with special needs and students who are learning English are incorporated into the materials and strategies used with all students during the initial instruction phase. In addition, the **Access and Equity** chapter in *Teacher Resources* (or go to FOSSweb to download this chapter) provides strategies and suggestions for enhancing the science and engineering experiences for each of the specific groups noted above.

Throughout the FOSS investigations, students experience multiple ways of interacting with phenomena and expressing their understanding through a variety of modalities. Each student has multiple opportunities to demonstrate his or her strengths and needs, thoughts, and aspirations.

The challenge is then to provide appropriate follow-up experiences or enhancements appropriate for each student. For some students, this might mean more time with the active investigations or online activities. For other students, it might mean more experience and/or scaffolds for developing models, building explanations, or engaging in argument from evidence.

For some students, it might mean making vocabulary and language structures more explicit through new concrete experiences or through reading to students. It may help them identify and understand relationships and connections through graphic organizers.

For other students, it might be designing individual projects or small-group investigations. It might be more opportunities for experiencing science outside the classroom in more natural, outdoor environments or defining problems and designing solutions in their communities.

English Learners

The FOSS Program provides a rich laboratory for language development for English learners. A variety of techniques are provided to make science concepts clear and concrete, including modeling, visuals, and active investigations in small groups. Instruction is guided and scaffolded through carefully designed lesson plans, and students are supported throughout.

Science vocabulary and language structures are introduced in authentic contexts while students engage in hands-on learning and collaborative discussion. Strategies for helping all students read, write, speak, and listen are described in the Science-Centered Language Development chapter. A specific section on English learners provides suggestions for both integrating English language development (ELD) approaches during the investigation and for developing designated (targeted and strategic) ELD-focused lessons that support science learning.

FOSS INVESTIGATION ORGANIZATION

Modules are subdivided into **investigations** (four in this module). Investigations are further subdivided into three to five **parts**. Each investigation has a general guiding question for the phenomenon students investigate, and each part of each investigation is driven by a specific **focus question**. The focus question, usually presented as the part begins, engages the student with the phenomenon and signals the challenge to be met, mystery to be solved, or principle to be uncovered. The focus question guides students' actions and thinking and makes the learning goal of each part explicit for teachers. Each part concludes with students recording an answer to the focus question in their notebooks.

The investigation is summarized for the teacher in the At-a-Glance chart at the beginning of each investigation.

Investigation-specific **scientific background** information for the teacher is presented in each investigation chapter organized by the focus questions.

The **Teaching Children about** section makes direct connections to the NGSS foundation boxes for the grade level—Disciplinary Core Ideas, Science and Engineering Practices, and Crosscutting Concepts. This information is later presented in color-coded sidebar notes to identify specific places in the flow of the investigation where connections to the three dimensions of science learning appear. The Teaching Children about section ends with information about teaching and learning and a conceptual-flow graphic of the content.

The **Materials** and **Getting Ready** sections provide scheduling information and detail exactly how to prepare the materials and resources for conducting the investigation.

Teaching notes and **ELA Connections** appear in blue boxes in the sidebars. These notes comprise a second voice in the curriculum—an educative element. The first (traditional) voice is the message you deliver to students. The second (educative) voice, shared as a teaching note, is designed to help you understand the science content and pedagogical rationale at work behind the instructional scene. ELA Connections boxes provide connections to the Common Core State Standards for English Language Arts.

FOCUS QUESTION

What happens when you rub rocks together?

SCIENCE AND ENGINEERING PRACTICES

Planning and carrying out investigations

DISCIPLINARY CORE IDEAS

ESS2A: Earth materials and systems

CROSSCUTTING CONCEPTS

Stability and change

TEACHING NOTE

This focus question can be answered with a simple yes or no, but the question has power when students support their answers with evidence. Their answers should take the form "Yes, because _____."

The **Getting Ready** and **Guiding the Investigation** sections have several features that are flagged in the sidebars. These include several icons to remind you when a particular pedagogical method is suggested, as well as concise bits of information in several categories.

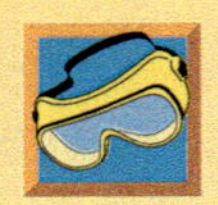

The **safety** icon alerts you to potential safety issues related to chemicals, allergic reactions, and the use of safety goggles.

The small-group **discussion** icon asks you to pause while students discuss data or construct explanations in their groups.

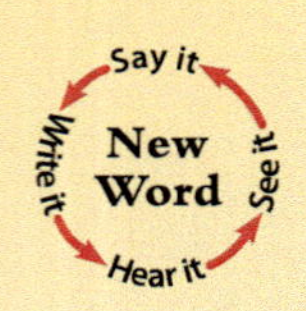

The **new-word** icon alerts you to a new vocabulary word or phrase that should be introduced thoughtfully.

The **vocabulary** icon indicates where students should review recently introduced vocabulary.

The **recording** icon points out where students should make a science-notebook entry.

The **reading** icon signals when the class should read a specific article in the *FOSS Science Resources* book.

The **technology** icon signals when the class should use a digital resource on FOSSweb.

The **assessment** icons appear when there is an opportunity to assess student progress by using embedded or benchmark assessments. Some are performance assessments—observations of science and engineering practices, indicated by a second icon which includes a beaker and ruler.

The **outdoor** icon signals when to move the science learning experience into the schoolyard.

The **engineering** icon indicates opportunities for an experience incorporating engineering practices.

The **math** icon indicates an opportunity to engage in numerical data analysis and mathematics practice.

The **crosscutting concepts** icon indicates an opportunity to expand on the concept by going to *Teacher Resources*, Crosscutting Concepts chapter. This chapter provides details on how to engage students with that concept in the context of the investigation.

The **EL note** provides a specific strategy to use to assist English learners in developing science concepts.

EL NOTE

To help with pacing, you will see icons for **breakpoints**. Some breakpoints are essential, and others are optional.

ESTABLISHING A CLASSROOM CULTURE

Part of being a scientist is learning how to work collaboratively with others. However, students in primary grades are usually most comfortable working as individuals with materials. The abilities to share, take turns, and learn by contributing to a group goal are developing but are not as reliable as learning strategies all the time. Because of this egocentrism and the need for many students to control materials or dominate actions, the FOSS kit includes a lot of materials. To effectively manage students and materials, here are some suggestions.

My Responsibilities

I agree that I will...

- explain my ideas.
- listen to others and show that I am listening.
- ask questions when I am confused or can't hear.
- connect my ideas to others' (explain, add to, respectfully disagree).
- participate because all ideas lead to learning (speak loud and clear).

This poster is an example of student responsibilities that the class discussed and adopted as their norms.

Whole-Class Discussions

Introducing and wrapping up the center activities require you to work for brief periods with the whole class. FOSS suggests for these introductions and wrap-ups that you gather the class at the rug or other location in the classroom where students can sit comfortably in a large group.

At the beginning of the year, explain and discuss norms for sense-making discussions. You might start by together making a class poster with visuals to represent what it looks like, sounds like, and feels like when everyone is working and learning together. Model discussion protocols that give all students opportunities to speak and listen, such as think-pair-share. Review the norms before sense-making discussions, and leave time for reflecting on how well the group adhered to the norms. More strategies for developing oral discourse skills can be found in Sense-Making Discussions for Three-Dimensional Learning and the Science-Centered Language Development chapters in *Teacher Resources* on FOSSweb.

Collaborative Teaching and Learning

Collaborative learning requires a collective as well as individual growth mindset. A growth mindset is when people believe that their most basic abilities can be developed through dedication and hard work (see the research of Carol Dweck and her book *Mindset: The Psychology of Success*). As second-grade students learn to work together to make sense of phenomena and develop their inquiry and discourse skills, it's important to recognize and value their efforts to try new approaches, to share their ideas, and ask questions. Remind students that everyone in the classroom is a learner, and that learning happens when we try to figure things out. Here are a few ways to help students develop a growth mindset for science and engineering.

- **Praise effort, not right answers**. When students are successful at a task, provide positive feedback about their level of engagement and effort in the practices, e.g., the efforts they put into careful observations, how well they reported their observations, the relevancy of their questions, how well they connected or applied new concepts, and their use of new vocabulary, etc. Also, try to provide feedback that encourages students to continue to improve their learning and exploring, e.g., is there another way you could try? Have you thought about _____? Why do you think _____?
- **Foster and validate divergent thinking**. During sense-making discussions, continually emphasize how important it is to share emerging ideas and to be open to the ideas of others in order to build understanding. Model for students how you refine and revise your thinking based on new information. Make it clear to students that the point is not for them to show they have the right answer, but rather to help each other arrive at new understanding. Point out positive examples of students expressing and revising their ideas. For example, "Did you all notice how Carla changed her idea about _____? "

Establishing a classroom culture that supports three-dimensional teaching and learning centers on collaboration. Helping students to work together in pairs and small groups, and to adhere to norms for discussions, are ways to foster collaboration. These structures along with the expectations that students will be negotiating meaning together as a community of learners, creates a learning environment where students are compelled to work, think, and communicate like scientists and engineers to help one another learn.

Small-Group Centers

Some of the observations and investigations with earth materials can be conducted with small groups at a learning center. For example, making sand sculptures in Investigation 3, Part 3, could be conducted at a center. Limit the number of students at the center to six to ten at one time. When possible, each student will have his or her own equipment to work with. In some cases, students will have to share materials and equipment and make observations together. As one group at a time is working at the center on a FOSS activity, other students will be doing something else. Over the course of an hour or more, plan to rotate all students through the center, or allow the center to be a free-choice station.

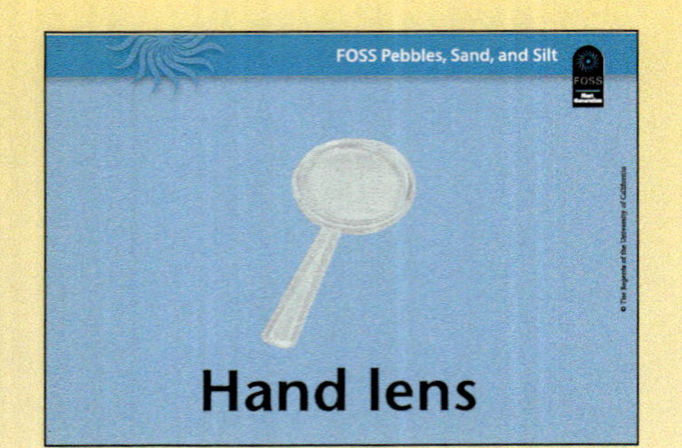

When You Don't Have Adult Helpers

Some parts of investigations work better when there is an aide or a student's family member available to assist groups with the activity and to encourage discussion and vocabulary development. We realize that there are many primary classrooms in which the teacher is the only adult present. You might invite upper-elementary students to visit your class to help with the activities. Remind older students to be guides and to let primary students do the activities themselves.

Managing Materials

The Materials section lists the items in the equipment kit and any teacher-supplied materials. It also describes things to do to prepare a new kit and how to check and prepare the kit for your classroom. Individual photos of each piece of FOSS equipment are available for printing from FOSSweb, and can help students and you identify each item. (Photo equipment cards are available in English and Spanish formats.)

For whole-class activities, FOSS Program designers suggest using a central materials distribution system. You organize all the materials for an investigation at a single location called the materials station. As the investigation progresses, one member of each group gets materials as they are needed, and another returns the materials when the investigation is complete. You place the equipment and resources at the station, and students do the rest. Students can also be involved in cleaning and organizing the materials at the end of a session.

When Students Are Absent

When a student is absent for a session, give him or her a chance to spend some time with the materials at a center. Another student might act as a peer tutor. Allow the student to bring home a *FOSS Science Resources* book to read with a family member. Each article has a few review items that the student can respond to verbally or in writing.

SAFETY IN THE CLASSROOM AND OUTDOORS

Following the procedures described in each investigation will make for a very safe experience in the classroom. You should also review your district safety guidelines and make sure that everything you do is consistent with those guidelines. Two posters are included in the kit: *Science Safety* for classroom use and *Outdoor Safety* for outdoor activities.

Look for the safety icon in the Getting Ready and Guiding the Investigation sections that will alert you to safety considerations throughout the module.

Safety Data Sheets (SDS) for materials used in the FOSS Program can be found on FOSSweb. If you have questions regarding any SDS, call Delta Education at 1-800-258-1302 (Monday–Friday, 8:00 a.m.–5:00 p.m. ET).

Science Safety in the Classroom

General classroom safety rules to share with students are listed here.

1. Listen carefully to your teacher's instructions. Follow all directions. Ask questions if you don't know what to do.
2. Tell your teacher if you have any allergies.
3. Never put any materials in your mouth. Do not taste anything unless your teacher tells you to do so.
4. Never smell any unknown material. If your teacher tells you to smell something, wave your hand over the material to bring the smell toward your nose.
5. Do not touch your face, mouth, ears, eyes, or nose while working with chemicals, plants, or animals.
6. Always protect your eyes. Wear safety goggles when necessary. Tell your teacher if you wear contact lenses.
7. Always wash your hands with soap and warm water after handling chemicals, plants, or animals.
8. Never mix any chemicals unless your teacher tells you to do so.
9. Report all spills, accidents, and injuries to your teacher.
10. Treat animals with respect, caution, and consideration.
11. Clean up your work space after each investigation.
12. Act responsibly during all science activities.

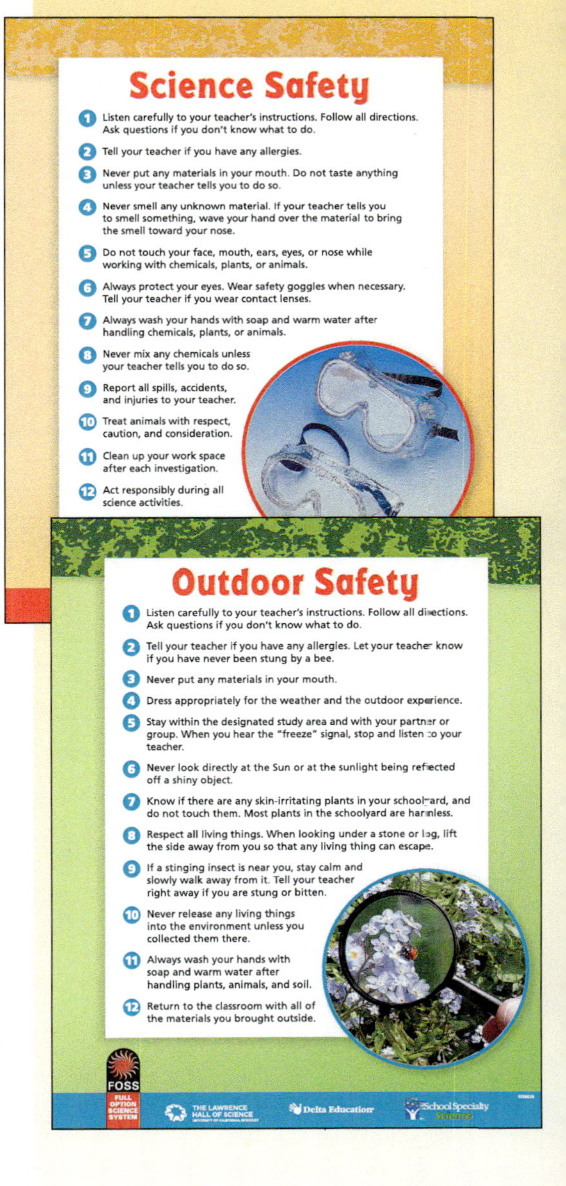

NOTE
The Getting Ready section for each part of an investigation helps you prepare. It provides information on scheduling the activities and introduces the tools and techniques used in the activity. Be prepared—read the Getting Ready section thoroughly and review the teacher preparation video on FOSSweb.

SCHEDULING THE MODULE

On the next page is a suggested teaching schedule for the module. The investigations are numbered and should be taught in order, as the concepts build upon each other from investigation to investigation. We suggest that a minimum of 9 weeks be devoted to this module.

Active-investigation (A) sessions include hands-on work with materials and tools, active thinking about experiences, small-group discussion, writing in science notebooks, and learning new vocabulary in context.

Reading (R) sessions involve reading *FOSS Science Resources* articles. Reading can be completed during language-arts time to make connections to Common Core State Standards for ELA (CCSS ELA).

During **Wrap-Up/Warm-Up (W)** sessions, students share notebook entries and engaging in connections to CCSS ELA. These sessions can also be completed during language-arts time.

I-Checks are short summative assessments at the end of each investigation. See the Assessment chapter for next-step strategies for self-assessment.

Week	Day 1	Day 2	Day 3	Day 4	Day 5
1	START Inv. 1 Part 1 A	A/W	START Inv. 1 Part 2 A	A/W	START Inv. 1 Part 3 A/W
2	START Inv. 1 Part 4 A	A	R/W	START Inv. 1 Part 5 A/R	I-Check 1
3	START Inv. 2 Part 1 A	A/W	START Inv. 2 Part 2 A	R/W	START Inv. 2 Part 3 A
4	A/W	START Inv. 2 Part 4 A	A/R	R	I-Check 2
5	START Inv. 3 Part 1 A	R/W	START Inv. 3 Part 2 A/W		START Inv. 3 Part 3 A/W
6	START Inv. 3 Part 4 A	A/W	START Inv. 3 Part 5 A	A	R
7	I-Check 3		START Inv. 4 Part 1 A	A	A/W
8	START Inv. 4 Part 2 A	A	A	R	R/W
9	START Inv. 4 Part 3 A/R	A/R	START Inv. 4 Part 4 A/R	A/R	I-Check 4

FOSS CONTACTS

General FOSS Program information

www.FOSSweb.com

www.DeltaEducation.com/FOSS

Contact the developers at the Lawrence Hall of Science

foss@berkeley.edu

Customer Service at Delta Education

www.DeltaEducation.com/contact.aspx

Phone: 1-800-258-1302, 8:00 a.m.–5:00 p.m. ET

FOSSmap (online component of FOSS assessment system)

http://FOSSmap.com/

FOSSweb account questions/access codes/help logging in

techsupport.science@schoolspecialty.com

Phone: 1-800-258-1302, 8:00 a.m.–5:00 p.m. ET

School Specialty online support

loginhelp@schoolspecialty.com

Phone: 1-800-513-2465, 8:30 a.m. –6:00 p.m. ET

FOSSweb tech support

support@fossweb.com

Professional development

www.FOSSweb.com/Professional-Development

Safety issues

www.DeltaEducation.com/SDS

Phone: 1-800-258-1302, 8:00 a.m.–5:00 p.m. ET

For chemical emergencies, contact Chemtrec 24 hours a day.

Phone: 1-800-424-9300

Sales and replacement parts

www.DeltaEducation.com/FOSS/buy

Phone: 1-800-338-5270, 8:00 a.m.–5:00 p.m. ET

Framework and NGSS

Framework and NGSS

INTRODUCTION TO PERFORMANCE EXPECTATIONS

"The NGSS are standards or goals, that reflect what a student should know and be able to do; they do not dictate the manner or methods by which the standards are taught. . . . Curriculum and assessment must be developed in a way that builds students' knowledge and ability toward the PEs [performance expectations]" (Next Generation Science Standards, 2013, page xiv).

This chapter shows how the NGSS Performance Expectations were bundled in the **Pebbles, Sand, and Silt Module** to provide a coherent set of instructional materials for teaching and learning.

This chapter also provides details about how this FOSS module fits into the matrix of the FOSS Program (page 38). Each FOSS module K–5 and middle school course 6–8 has a functional role in the FOSS conceptual frameworks that were developed based on a decade of research on science education and the influence of *A Framework for K–12 Science Education* (2012) and *Next Generation Science Standards* (NGSS, 2013).

The FOSS curriculum provides a coherent vision of science teaching and learning in the three ways described by the NRC *Framework*. First, FOSS is designed around learning as a developmental progression, providing experiences that allow students to continually build on their initial notions and develop more complex science and engineering knowledge. Students develop functional understanding over time by building on foundational elements (intermediate knowledge). That progression is detailed in the conceptual frameworks.

Second, FOSS limits the number of core ideas, choosing depth of knowledge over broad shallow coverage. Those core ideas are addressed at multiple grade levels in ever greater complexity. FOSS investigations at each grade level focus on elements of core ideas that are teachable and learnable at that grade level.

Third, FOSS investigations integrate engagement with scientific ideas (content) and the practices of science and engineering by providing firsthand experiences.

Teach the module with the confidence that the developers have carefully considered the latest research and have integrated into each investigation the three dimensions of the *Framework* and NGSS, and have designed powerful connections to the Common Core State Standards for English Language Arts.

Contents

The NGSS Performance Expectations bundled in this module include:

Earth and Space Sciences
2-ESS1-1
2-ESS2-1
2-ESS2-2
2-ESS2-3

Physical Sciences
2-PS1-1
2-PS1-2

Engineering, Technology, and Applications of Science
K–2 ETS1-1
K–2 ETS1-2
K–2 ETS1-3

DISCIPLINARY CORE IDEAS

A Framework for K–12 Science Education has three core ideas in Earth and space sciences.

ESS1: Earth's place in the universe

ESS2: Earth's systems

ESS3: Earth and human activity

Only two of these core ideas are in the NGSS performance expectations for grade 2.

The questions and descriptions of the core ideas in the text on these pages are taken from the NRC *Framework* for the grades K–2 grade band to keep the core ideas in a rich and useful context.

The performance expectations related to each core idea are taken from the NGSS for grade 2.

Disciplinary Core Ideas Addressed

The **Pebbles, Sand, and Silt Module** connects with the NRC *Framework* for the grades K–2 grade band and the NGSS performance expectations for grade 2. The module focuses on core ideas for earth science, matter, and engineering design.

Earth and Space Sciences

Framework core idea ESS1: Earth's place in the universe—What is the universe, and what is Earth's place in it?

- **ESS1.C: The history of planet Earth**
  How do people reconstruct and date events in Earth's planetary history? [Some events on Earth occur in cycles, like day and night, and other have a beginning and an end, like a volcanic eruption. Some events, like an earthquake, happen very quickly; others such as the formation of the Grand Canyon, occur very slowly, over a time period much longer than one can observe.]

The following NGSS Grade 2 Performance Expectation for ESS1 is derived from the Framework disciplinary core ideas above.

- **2-ESS1-1.** Make observations from media to construct an evidence-based account that Earth events can occur quickly or slowly. [Clarification Statement: Examples of events and timescales could include volcanic explosions and earthquakes, which happen quickly, and erosion of rocks, which occurs slowly.]

Framework core idea ESS2: Earth's systems—How and why is Earth constantly changing?

- **ESS2.A: Earth materials and systems**
  How do the major Earth systems interact? [Wind and water can change the shape of the land. The resulting landforms, together with the materials on the land, provide homes for living things.]
- **ESS2.B: Plate tectonics and large-scale system interactions**
  Why do the continents move, and what causes earthquakes and volcanoes? [Rocks, soils, and sand are present in most areas where plants and animals live. There may also be rivers, streams, lakes, and ponds. Maps show where things are located. One can map the shapes and kinds of land and water in any area.]

- **ESS2.C: The roles of water in Earth's surface processes**
  How do the properties and movements of water shape Earth's surface and affect its systems? [Water is found in the ocean, rivers, lakes, and ponds. Water exists as solid ice and in liquid form. It carries soil and rocks from one place to another and determines the variety of life forms that can live in a particular location.]

The following NGSS Grade 2 Performance Expectations for ESS2 are derived from the Framework disciplinary core ideas above.

- **2-ESS2-1**. Compare multiple solutions designed to slow or prevent wind or water from changing the shape of the land. [Clarification Statement: Examples of solutions could include different designs of dikes and windbreaks to hold back wind and water, and different designs for using shrubs, grass, and trees to hold back the land.]
- **2-ESS2-2.** Develop a model to represent the shapes and kinds of land and bodies of water in an area. [Assessment Boundary: Assessment does not include quantitative scaling in models.]
- **2-ESS2-3.** Obtain information to identify where water is found on Earth and that it can be solid or liquid.

REFERENCES

National Research Council. *A Framework for K–12 Science Education: Practices, Crosscutting Concepts, and Core Ideas.* Washington, DC: National Academies Press, 2012.

NGSS Lead States. *Next Generation Science Standards: For States, by States.* Washington, DC: National Academies Press, 2013.

National Governors Association Center for Best Practices and Council of Chief State School Officers. *Common Core State Standards for English Language Arts and Literacy in History/Social Studies, Science, and Technical Subjects, Washington, DC:* 2010.

DISCIPLINARY CORE IDEAS

A Framework for K–12 Science Education has four core ideas in physical sciences.

- PS1: Matter and its interactions
- PS2: Motion and stability: Forces and interactions
- PS3: Energy
- PS4: Waves and their applications in technologies for information transfer

The questions and descriptions of the core ideas in the text on these pages are taken from the NRC *Framework* for the grades K–2 grade band to keep the core ideas in a rich and useful context.

The performance expectations related to each core idea are taken from the NGSS for grade 2.

Physical Sciences

Framework core idea PS1: Matter and its interactions—How can one explain the structure, properties, and interactions of matter?

- **PS1.A: Structure and properties of matter**
  How do particles combine to form the variety of matter one observes? [Different kinds of matter exist (e.g., wood, metal, water), and many of them can be either solid or liquid, depending on temperature. Matter can be described and classified by its observable properties, by its uses, and by whether it occurs naturally or is manufactured. Different properties are suited to different purposes. A great variety of objects can be built up from a small set of pieces. Objects or samples of a substance can be weighed, and their size can be described and measured. (Boundary: Volume is introduced only for liquid measure.)]

The following NGSS Grade 2 Performance Expectations for PS1 are derived from the Framework disciplinary core ideas above.

- **2-PS1-1.** Plan and conduct an investigation to describe and classify different kinds of materials by their observable properties. [Clarification Statement: Observations could include color, texture, hardness, and flexibility. Patterns could include the similar properties that different materials share.]
- **2-PS1-2.** Analyze data obtained from testing different materials to determine which materials have the properties that are best suited for an intended purpose. [Clarification Statement: Examples of properties could include, strength, flexibility, hardness, texture, and absorbency.]

Engineering, Technology, and Applications of Science

Framework core idea ETS1: Engineering design—How do engineers solve problems?

- **ETS1.A: Defining and delimiting an engineering problem**
  What is a design for? What are the criteria and constraints of a successful solution? [A situation that people want to change or create can be approached as a problem to be solved through engineering. Such problems may have many acceptable solutions. Asking questions, making observations, and gathering information are helpful in thinking about problems. Before beginning to design a solution, it is important to clearly understand the problem.]
- **ETS1.B: Developing possible solutions**
  What is the process for developing potential design solutions? [Designs can be conveyed through sketches, drawings or physical models. These representations are useful in communicating ideas for a problem's solutions to other people. To design something complicated, one may need to break the problem into parts and attend to each part separately but must then bring the parts together to test the overall plan.]
- **ETS1.C: Optimizing the design solution**
  How can the various proposed design solutions be compared and improved? [Because there is always more than one possible solution to a problem, it is useful to compare designs, test them, and discuss their strengths and weakness.]

The following NGSS Grades K–2 Performance Expectations for ETS1 are derived from the Framework disciplinary core ideas above.

- **K-2-ETS1-1.** Ask questions, make observations, and gather information about a situation people want to change to define a simple problem that can be solved through the development of a new or improved object or tool.
- **K-2-ETS1-2.** Develop a simple sketch, drawing, or physical model to illustrate how the shape of an object helps it function as needed to solve a given problem.
- **K-2-ETS1-3.** Analyze data from tests of two objects designed to solve the same problem to compare the strengths and weaknesses of how each performs.

DISCIPLINARY CORE IDEAS

A Framework for K–12 Science Education has two core ideas in engineering, technology, and applications of science.

ETS1: Engineering design

ETS2: Links among engineering, technology, science, and society

NOTE: Only one of these core ideas, ETS1, is represented in the NGSS performance expectations for grade 2.

The questions and descriptions of the core ideas in the text on these pages are taken from the NRC *Framework* for the grades K–2 grade band to keep the core ideas in a rich and useful context.

The performance expectations related to each core idea are taken from the NGSS for grade K–2.

Framework core idea ETS2: Links among engineering, technology, science, and society—How are engineering, technology, science, and society interconnected?

- **ETS2.A: Interdependence of science, engineering, and technology**
  What are the relationships among science, engineering, and technology? [People encounter questions about the natural world every day. There are many types of tools produced by engineering that can be used in science to help answer these questions through observation or measurement. Observations and measurements are also used in engineering to help test and refine design ideas.]
- **ETS2.B: Influence of engineering, technology, and science on society and the natural world**
  How do science, engineering, and the technologies that result from them affect the ways in which people live? How do they affect the natural world? [People depend on various technologies in their lives; human life would be very different without technology. Every human-made product is designed by applying some knowledge of the natural world and is built by using materials derived from the natural world, even when the materials are not themselves natural—for example, spoons made from refined metals. Thus, developing and using technology has impacts on the natural world.]

Note: There are no separate performance expectations described for core idea ETS2 (see volume 2, appendix J, for an explanation and elaboration).

Science and Engineering Practices Addressed

1. ***Asking questions and defining problems***
   - Ask questions based on observations to find more information about the natural and/or designed world(s).
   - Define a simple problem that can be solved through the development of a new or improved object or tool.
2. ***Developing and using models***
   - Distinguish between a model and the actual object, process, and/or events the model represents.
   - Compare models to identify common features and differences.
   - Develop and/or use a model to represent amounts, relationships, relative scales, and/or patterns in the natural world.
   - Develop a simple model based on evidence to represent a proposed object or tool.
3. ***Planning and carrying out investigations***
   - Plan and conduct an investigation collaboratively to produce data to serve as the basis for evidence to answer a question.
   - Make observations (firsthand or from media) and/or measurements to collect data that can be used to make comparisons.
   - Make predictions based on prior experiences.
4. ***Analyzing and interpreting data***
   - Record information (observations, thoughts, and ideas).
   - Use and share pictures, drawings, and/or writings of observations.
   - Use observations (firsthand or from media) to describe patterns and/or use relationships in the natural and designed world(s) in order to answer scientific questions and solve problems.
   - Analyze data from tests of an object or tool to determine if it works as intended.
5. ***Using mathematics and computational thinking***
   - Describe, measure, and/or compare quantitative attributes of different objects and display the data using simple graphs.
6. ***Constructing explanations and designing solutions***
   - Make observations (firsthand or from media) to construct an evidence-based account for natural phenomena.
   - Compare multiple solutions to a problem.

SCIENCE AND ENGINEERING PRACTICES

A Framework for K–12 Science Education (National Research Council, 2012) describes eight science and engineering practices as essential elements of a K–12 science and engineering curriculum. All eight practices are incorporated into the learning experiences in the **Pebbles, Sand, and Silt Module**.

The learning progression for this dimension of the framework is addressed in *Next Generation Science Standards* (National Academies Press, 2013), volume 2, appendix F. Elements of the learning progression for practices recommended for grade 2 as described in the performance expectations appear in bullets below each practice.

CROSSCUTTING CONCEPTS

A Framework for K–12 Science Education describes seven crosscutting concepts as essential elements of a K–12 science and engineering curriculum. The crosscutting concepts listed here include those recommended for grade 2 in the NGSS and are incorporated into the learning opportunities in the **Pebbles, Sand, and Silt Module**.

The learning progression for this dimension of the framework is addressed in volume 2, appendix G, in the NGSS. Elements of the learning progression for crosscutting concepts recommended for grade 2, as described in the performance expectations, appear after bullets below each concept.

7. Engaging in argument from evidence

- Construct an argument with evidence to support a claim.
- Make a claim about the effectiveness of an object, tool, or solution that is supported by relevant evidence.

8. Obtaining, evaluating, and communicating information

- Read grade-appropriate texts and/or use media to obtain scientific and/or technical information to determine patterns in and/or evidence about the natural and designed world(s).
- Obtain information using various texts, text features (e.g., headings, tables of contents, glossaries, electronic menus, icons), and other media that will be useful in answering a scientific question.
- Communicate information or design ideas and/or solutions with others in oral and/or written forms using models, drawings, writing, or numbers that provide detail about scientific ideas, practices, and/or design ideas.

Crosscutting Concepts Addressed

Patterns

- Patterns in the natural and human designed world can be observed, used to describe phenomena, and used as evidence.

Cause and effect

- Events have causes that generate observable patterns.

Scale, proportion, and quantity

- Relative scales allow objects and events to be compared and described.

Energy and matter

- Objects may break into smaller pieces, be put together into larger pieces, or change shapes.

Stability and change

- The shape and stability of structures of natural and designed objects are related to their function(s).

Connections: Understandings about the Nature of Science

Scientific investigations use a variety of methods.

- Scientific investigation begin with a question. Scientists use different ways to study the world.

Scientific knowledge is based on empirical evidence.

- Scientists look for patterns and order when making observations about the world.
- Many events are repeated.

Science is a way of knowing.

- Science knowledge informs us about the world.

Science addresses questions about the natural and material world.

- Scientists study the natural and material world.

Science is a human endeavor.

- Science effects everyday life.

Connections: Engineering, Technology, and Application of Science

Influence of engineering, technology, and science on society and the natural world.

- Every human-made product is designed by applying some knowledge of the natural world and is built by using natural materials.

CONNECTIONS

See volume 2, appendix H and appendix J, in the NGSS for more on these connections.

For details on learning connections to Common Core State Standards English Language Arts and Math, see the chapters FOSS and Common Core ELA—Grade 2 and FOSS and Common Core Math—Grade 2 in *Teacher Resources*.

FOSS CONCEPTUAL FRAMEWORK

In the last half decade, teaching and learning research has focused on learning progressions. The idea behind a learning progression is that **core ideas** in science are complex and wide-reaching, requiring years to develop fully—ideas such as the structure of matter or the relationship between the structure and function of organisms. From the age of awareness throughout life, matter and organisms are important to us. There are things students can and should understand about these core ideas in primary school years, and progressively more complex and sophisticated things they should know as they gain experience and develop cognitive abilities. When we as educators can determine those logical progressions, we can develop meaningful and effective curriculum for students.

FOSS has elaborated learning progressions for core ideas in science for kindergarten through grade 8. Developing a learning progression involves identifying successively more sophisticated ways of thinking about a core idea over multiple years.

If mastery of a core idea in a science discipline is the ultimate educational destination, then well-designed learning progressions provide a map of the routes that can be taken to reach that destination. . . . Because learning progressions extend over multiple years, they can prompt educators to consider how topics are presented at each grade level so that they build on prior understanding and can support increasingly sophisticated learning. (National Research Council, *A Framework for K–12 Science Education*, 2012, p. 26)

TEACHING NOTE

FOSS has conceptual structure at the module and strand levels. The concepts are carefully selected and organized in a sequence that makes sense to students when presented as intended.

The FOSS modules are organized into three domains: physical science, earth science, and life science. Each domain is divided into two strands, as shown in the table "FOSS Next Generation—K–8 Sequence." Each strand represents a core idea in science and has a conceptual framework.

- Physical Science: matter; energy and change
- Earth and Space Science: dynamic atmosphere; rocks and landforms
- Life Science: structure and function; complex systems

The sequence in each strand relates to the core ideas described in the NRC *Framework*. Modules at the bottom of the table form the foundation in the primary grades. The core ideas develop in complexity as you proceed up the columns.

Information about the FOSS learning progression appears in the **conceptual framework** (page 41, 43), which shows the structure of scientific knowledge taught and assessed in this module, and the **content sequence** (pages 44–45), a graphic and narrative description that puts this single module into a K–8 strand progression.

FOSS is a research-based curriculum designed around the core ideas described in the NRC *Framework*. The FOSS module sequence provides opportunities for students to develop understanding over time by building on foundational elements or intermediate knowledge leading to the understanding of core ideas. Students develop this understanding by engaging in appropriate science and engineering practices and exposure to crosscutting concepts. The FOSS conceptual frameworks therefore are *more detailed* and *finer-grained* than the set of goals described by the NGSS performance expectations (PEs). The following statement reinforces the difference between the standards as a blueprint for assessment and a curriculum, such as FOSS.

FOSS Next Generation—K–8 Sequence

	PHYSICAL SCIENCE		EARTH SCIENCE		LIFE SCIENCE	
	MATTER	ENERGY AND CHANGE	ATMOSPHERE AND EARTH	ROCKS AND LANDFORMS	STRUCTURE/ FUNCTION	COMPLEX SYSTEMS
6–8	Waves; Gravity and Kinetic Energy Chemical Interactions Electromagnetic Force; Variables and Design		Planetary Science Earth History Weather and Water		Heredity and Adaptation Human Systems Interactions; Populations and Ecosystems Diversity of Life	
5	Mixtures and Solutions		Earth and Sun		Living Systems	
4		Energy		Soils, Rocks, and Landforms	Environments	
3	Motion and Matter		Water and Climate		Structures of Life	
2	Solids and Liquids			Pebbles, Sand, and Silt	Insects and Plants	
1		Sound and Light	Air and Weather		Plants and Animals	
K	Materials and Motion		Trees and Weather		Animals Two by Two	

Some reviewers of both public drafts [of NGSS] *requested that the standards specify the intermediate knowledge necessary for scaffolding toward eventual student outcomes. However, the NGSS are a set of goals. They are PEs for the end of instruction—not a curriculum. Many different methods and examples could be used to help support student understanding of the DCIs and science and engineering practices, and the writers did not want to prescribe any curriculum or constrain any instruction. It is therefore outside the scope of the standards to specify intermediate knowledge and instructional steps.* (*Next Generation Science Standards*, 2013, volume 2, p. 342)

BACKGROUND FOR THE CONCEPTUAL FRAMEWORK *in Pebbles, Sand, and Silt*

There are two conceptual frameworks for the disciplinary core ideas in this module for grade 2—one with a focus on earth and physical science dealing with structures and interactions of earth materials and one on engineering design.

Studying Rocks

The oldest rocks on Earth we know about are approximately 3.8 to 4.3 billion years old. They were found on the shores of Hudson Bay in Northern Canada. Some of the mineral grains (zircon) found in Western Australia have been dated to an age of 4.4 billion years old. That makes rocks just about the oldest things you can collect (the oldest objects yet discovered are meteorites, 4.5 to 4.6 billion years old).

Rocks have been important to humans in many ways for thousands of years. Rocks provided early people with shelter, weapons, and a means for creating sparks to start fires for cooking and warmth. The ancient Egyptians built their pyramids out of rock, and the Greeks used stones called calculi for adding and subtracting. Hopi women used flat stones of different roughnesses to grind corn. Children during America's colonial days wrote on flat pieces of rock called slate. Astronauts collected rocks during their visits to the Moon. They brought these rocks back to scientists who hoped to unlock the secrets of the solar system by studying the rocks.

Rocks are the solid earth materials that compose the bulk of this planet. There are thousands of kinds of rocks—differences in their properties are what geologists use to recognize and distinguish them. There are a variety of methods and tools, but a geologist's first observations are very similar to the ones students will make in this module. The macroscopic properties (color, texture, and grain size) observed in the field with the naked eye or a 10X hand lens, help geologists begin classifying the rocks. Back in the lab, more refined techniques can be applied that define the rock's composition, age, and history.

The size of the rock grains or particles provides important clues to the history of a rock. There are several particle-size classification systems. The Wentworth scale is the one modified for these investigations.

Other scales have been devised by soil scientists, construction engineers, and the companies that collect, separate, and sell building materials. Most scales use the same terms to describe the particle sizes (e.g., pebble, gravel, sand) with little variance in size.

Soil scientists have developed some field tests that help identify the smaller particle sizes. Sand feels gritty when you rub it between your fingers. Silt feels rough but not gritty. Clay feels greasy or slick. Any combination of these particle sizes may be found in soil. Soil scientists become quite adept at estimating the amounts of each size in a ball of soil by squeezing a lump, or cast, in their hands. The sizes of material in the cast are judged by how easily it breaks or crumbles.

The kinds of rocks and types of soils vary from one place to another around the world, and even from neighborhood to neighborhood. But no matter where a rock or soil is found, the first steps toward studying it include the same types of observations made in these activities.

Minerals and Rocks

According to one estimate, over 4,000 different minerals have been identified in Earth's crust, and new ones are still being discovered. Minerals are the basic ingredients that make up the crust. Minerals are chemical compounds that occur naturally, and their composition is expressed by a chemical formula, such as NaCl, salt.

Minerals may occur as deposits of pure materials, or they may be found in combination with other minerals, forming rocks. Each mineral in a rock has its own identifiable physical and chemical properties, which contribute to the properties of the rock.

Soil

What is soil? To the farmer, soil is the layer of earth material in which plants anchor their roots and from which they get the nutrients and water they need to grow. To a geologist, soil is the layer of earth materials at Earth's surface that has been produced by weathering of rocks and sediments and that hasn't moved from its original location. To an engineer, soil is any ground that can be dug up by earth-moving equipment and requires no blasting.

CONCEPTUAL FRAMEWORK
Earth Science, Rocks and Landforms: Pebbles, Sand, and Silt

Structure of Earth

Concept A The geosphere (lithosphere) has properties that can be observed and quantified.

- Rocks are the solid material of Earth and can be described by their properties. Rocks are composed of minerals.
- Rocks can be sorted into different sizes that include clay, silt, sand, gravel, pebbles, cobbles, and boulders.
- Soil is made partly from weathered rock and partly from organic materials. Soils vary from place to place and differ in their ability to support plants.
- Landforms and bodies of water can be represented in models and maps.

Concept B The hydrosphere has properties that can be observed and quantified.

- Water exists in three states on Earth: solid, liquid, and gas.
- Sources of water can be fresh (rivers, lakes, ponds) or salt water (ocean).

Concept C Humans depend on Earth's land, ocean, atmosphere, and biosphere for many different resources.

- People use earth materials to make and construct things. The properties of different earth materials make them suitable for specific uses.

Earth Interactions

Concept A All Earth processes are the result of energy flowing and matter cycling within and among the planet's systems.

- Wind and water can change the shape of the land.
- Weathering, caused by wind or water, causes larger rocks (boulders, cobbles) to break into smaller rocks (pebbles, gravel, sand, silt, clay).
- Some Earth events happen very quickly; others occur very slowly over a time period much longer than one can observe.

And to students, soil is dirt. In FOSS, *soil* is defined as a mixture of different-sized earth materials, such as gravel, sand, and silt, and an organic material called humus. Humus is the dark, musty-smelling stuff derived from the decayed and decomposed remains of plant and animal life. The proportions of these materials that make up soil differ from one location to another. All life depends on a dozen or so elements that must ultimately be derived from Earth's crust. Soil has been called the bridge between earth material and life; only after minerals have been broken down and incorporated into the soil can plants process the nutrients and make them available to people and other animals.

Water

Water is so common and familiar that we usually don't think about it as something to be defined or described. To the biologist, water is the sanctuary in which life was born. The complex chemistry of life was then, as it is today, water based. To the geologist, water is a liquid earth material, one of the substances (along with solid rocks and minerals and atmospheric gases) that make up or come from Earth. Water plays a central role in sculpting the planet's surface and in causing and moderating Earth's weather and climate.

On planet Earth, water is found naturally everywhere—in puddles, ponds, streams, and in the ocean. Even though water covers nearly three quarters of our planet's surface—hardly scarce—it is still the most precious substance on Earth. The ocean contains more than 97% of Earth's supply of water. But because water has run through, washed over, and worn down the rocky crust for billions of years, the seas have become repositories for a huge burden of dissolved minerals, mostly salts. Although salt water is the one and only environment for thousands of life-forms on Earth, salt water will not support human life. Less than 3% of Earth's water is fresh, and much of this fresh water is not readily accessible The water in rivers and lakes is easy to scoop up to quench a thirst, but water in glaciers and the atmosphere is harder to get for human use.

Water has some unique properties. It is the only material that occurs naturally on Earth's surface in all three states of matter: solid, liquid, and gas (water vapor). Earth's temperature range, allied with the particular properties of water, allows all three states (forms) to exist.

Engineering Design

Science is a discovery activity, a process for producing new knowledge. Scientific knowledge advances when scientists observe objects and events, think about how their observations related to what is known, test their ideas in logical ways, and generate explanations that integrate the new information into understanding of the natural world. Thus the scientific enterprise is both what we know (content knowledge) and how we come to know it (practices). The practices that engineers use are very similar to science practices but also involve defining problems and designing solutions.

The process of engineering design, while it involves engineering practices, is considered a separate set of disciplinary core ideas in the *Framework* and in the NGSS. There are three basic ideas of engineering design: defining the problem, developing possible solutions, and comparing solutions to improve the design.

Defining the problem involves asking questions and making observations to obtain information about designing a specific structure.

Developing possible solutions involves making decisions about the materials available and thoughtfully making a design about how they can be used. They should communicate their solutions orally and with drawings and words.

Comparing different solutions involves testing several designs to see how well each one meets the challenge. Second graders are not expected to conduct tests with controlled variables, but they should be able to determine if the structure meets the challenge and if not, how it might be improved. Collaboration is an important aspect of engineering design; learning from the successes and failures of other design groups can be very productive. Students can engage in engineering practices without fully engaging in the iterative process of design.

In this module, students explore the disciplinary core ideas of engineering design in the context of investigating how earth materials are used by people and how to optimize the design solution. FOSS has a continuum of engagements in the engineering practices and process from short experiences to more in-depth experiences where students reflect on the core ideas about the design process.

CONCEPTUAL FRAMEWORK
Engineering Design: Pebbles, Sand, and Silt

Concept A Defining and delimiting engineering problems.

Asking questions, making observations, and gathering information are helpful in thinking about a problem. Before beginning to design a solution, it is important to clearly understand the problem.

Concept B Developing possible solutions.

Designs can be conveyed through sketches, drawings, or physical models. These representations are useful in communicating ideas for a problem's solutions to other people.

Concept C Optimizing the design solution.

Because there is always more than one possible solution to a problem, it is useful to compare and test designs.

Rocks and Landforms Content Sequence

This table shows the five FOSS modules and courses that address the content sequence "rocks and landforms" for grades K–8. Running through the sequence are the two progressions—structure of Earth and Earth interactions. The supporting elements in each module (somewhat abbreviated) are listed. The elements for the **Pebbles, Sand, and Silt Module** are expanded to show how they fit into the sequence.

	ROCKS AND LANDFORMS	
Module or course	**Structure of Earth**	**Earth interactions**
Earth History	• The geological time scale, interpreted from rock strata and fossils, provides a way to organize Earth's history. Lower layers are older than higher layers—superposition. • Earth's crust is fractured into plates that move over, under, and past one another. • Volcanoes and earthquakes occur along plate boundaries. • The rock cycle is a way to describe the process by which new rock is created.	• Landforms are shaped by slow, persistent processes driven by weathering, erosion, deposition, and plate tectonics. • Water's movement changes Earth's surface. • Energy is derived from the Sun and Earth's hot interior. • All Earth processes are the result of energy flowing and matter cycling within and among Earth's systems. • Evolution is shaped by geological conditions.
Soils, Rocks, and Landforms	• Soils are composed of different kinds and amounts of earth materials (sediments) and humus; they can be described by their properties. • Rocks are made of minerals; rocks and minerals can be described and identified by their properties: hardness, streak, luster, and cleavage. • Earth materials are natural resources.	• Physical and chemical weathering breaks rock into smaller pieces (sediments). • Erosion is the movement of sediments; deposition is the process by which sediments come to rest in another place. • Landslides, earthquakes, and volcanoes can produce significant changes in landforms in a short period of time. • Downhill movement of water shapes land.
Water and Climate	• Water is found almost everywhere on Earth (e.g., vapor, clouds, rain, snow, ice). Most of Earth's water is in the ocean. • Water expands when heated, contracts when cooled, and expands when it freezes. • Cold water is more dense than warmer water; liquid water is more dense than ice. • Scientists observe, measure, and record patterns of weather to make predictions. • Soils retain more water than rock particles alone.	• Water flows downhill. • Ice melts when heated; water freezes when cooled. • The water cycle is driven by the Sun and involves evaporation, condensation, precipitation, and runoff. • Density determines whether objects float or sink in water. • Climate is the range of an area's typical weather. • A variety of natural hazards result from weather-related phenomena.
Pebbles, Sand, and Silt		
Materials and Motion	• Land, air, water, and trees are natural resources.	• Water interacts with natural and human-made materials.

NOTE

See the Assessment chapter at the end of this *Investigations Guide* for more details on how the FOSS embedded and benchmark assessment opportunities align to the conceptual frameworks and the learning progressions. In addition, the Assessment chapter describes specific connections between the FOSS assessments and the NGSS performance expectations.

The NGSS Performance Expections addressed in this module include:

Earth and Space Sciences
2-ESS1-1
2-ESS2-1
2-ESS2-2
2-ESS2-3

Physical Sciences
2-PS1-1
2-PS1-2

Engineering, Technology, and Applications of Science
K–2 ETS1-1
K–2 ETS1-2
K–2 ETS1-3

See pages 30–34 in this chapter for more details on the Grade 2 NGSS Performance Expectations.

Pebbles, Sand, and Silt

Structure of Earth	Earth interactions
• Rocks are earth materials composed of minerals; rocks can be described by their properties. • Rock sizes include clay, silt, sand, gravel, pebbles, cobbles, and boulders. • The properties of different earth materials (natural resources) make each suitable for specific uses. • Natural sources of water include streams, rivers, ponds, lakes, marshes, and the ocean. Sources of water can be fresh or salt water. • Water can be a solid, liquid, or gas. • Landforms and bodies of water can be represented in models and maps.	• Smaller rocks (sand) result from the breaking (weathering) of larger rocks. • Water carries soils and rocks from one place to another—erosion. • Some Earth events happen very quickly; others occur very slowly. • Wind and water can change the shape of the land. • Soil is made partly from weathered rock and partly from organic material. • Soils vary from place to place. Soils differ in their ability to support plants. • Earth materials are commonly used in the construction of buildings and streets.

CONNECTIONS TO NGSS BY INVESTIGATION

	Science and Engineering Practices	Connections to Common Core State Standards—ELA
Inv. 1: First Rocks	Asking questions Planning and carrying out investigations Analyzing and interpreting data Constructing explanations Engaging in argument from evidence Obtaining, evaluating, and communicating information	RI 1: Ask and answer questions to demonstrate understanding. RI 4: Determine the meaning of words and phrases in the text. RI 5: Know and use text features. RI 6: Identify the main purpose of the text. RI 7: Explain how images contribute to and clarify text. RI 9: Compare and contrast two texts on the same topic. W 5: Strengthen writing by revising and editing. W 7: Record science observations. SL 1: Participate in collaborative conversations. SL 2: Recount or describe key ideas.
Inv. 2: River Rocks	Developing and using models Planning and carrying out investigations Analyzing and interpreting data Using mathematics and computational thinking Constructing explanations Engaging in argument from evidence Obtaining, evaluating, and communicating information	RI 1: Ask and answer questions to demonstrate understanding. RI 2: Identify the main topic of the text. RI 3: Describe the connection between scientific ideas or concepts. RI 8: Describe how reasons support points the author makes in the text. RF 4: Read with accuracy and fluency to support comprehension. W 3: Write narratives. W 8: Recall information from experiences or gather information from provided sources to answer a question. SL 1: Participate in collaborative conversations. SL 2: Recount or describe key ideas. SL 4: Recount an experience. SL 5: Add drawings or other visual displays to recounts of experiences. L 4: Determine or clarify the meaning of unknown or multiple-meaning words and phrases. L 6: Use acquired words and phrases.

Disciplinary Core Ideas		Crosscutting Concepts
ESS1.C: The history of planet Earth • Some events happen very quickly; others occur very slowly over a time period much longer than one can observe. **(2-ESS1-1)**	**PS1.A: Structure and properties of matter** • Different kinds of matter exist and many of them can be either solid or liquid, depending on temperature. Matter can be described and classified by its observable properties. **(2-PS1-1)**	Patterns Cause and effect Stability and change
ESS1.C: The history of planet Earth • Some events happen very quickly; others occur very slowly over a time period much longer than one can observe. **(2-ESS1-1)** **ESS2.A: Earth materials and systems** • Wind and water can change the shape of the land. **(2-ESS2-1)** **ESS2.B: Plate tectonics and large-scale system interactions** • Maps show where things are located. One can map the shapes and kinds of land and water in any area. **(2-ESS2-2)** **ESS2.C: The roles of water in Earth's surface processes** • Water is found in the ocean, rivers, lakes, and ponds. Water exists as solid ice and in liquid form. **(2-ESS2-3)**	**PS1.A: Structure and properties of matter** • Different kinds of matter exist and many of them can be either solid or liquid, depending on temperature. Matter can be described and classified by its observable properties. **(2-PS1-1)**	Patterns Cause and effect Scale, proportion, and quantity Stability and change

	Science and Engineering Practices	Connections to Common Core State Standards—ELA
Inv. 3: Using Rocks	Defining problems Planning and carrying out investigations Analyzing and interpreting data Constructing explanations Engaging in argument from evidence Obtaining, evaluating, and communicating information	RI 1: Ask and answer questions to demonstrate understanding. RI 3: Describe the connection between scientific ideas or concepts. RI 5: Know and use text features. RF 4: Read with accuracy and fluency to support comprehension. SL 1: Participate in collaborative conversations. L 4: Determine or clarify the meaning of unknown or multiple-meaning words and phrases. L 5: Demonstrate understanding of word relationships and nuances in word meanings.
Inv. 4: Soil and Water	Asking questions and defining problems Developing and using models Planning and carrying out investigations Analyzing and interpreting data Constructing explanations and designing solutions Engaging in argument from evidence Obtaining, evaluating, and communicating information	RI 1: Ask and answer questions to demonstrate understanding. RI 2: Identify the main topic of the text. RI 3: Describe the connection between scientific ideas or concepts. RI 5: Know and use text features. RI 6: Identify the main purpose of the text. RI 7: Explain how images contribute to and clarify text. RF 4: Read with accuracy and fluency to support comprehension. W 7: Participate in shared research and writing projects. SL 1: Participate in collaborative conversations. SL 2: Recount or describe key ideas. SL 3: Ask and answer questions. SL 4: Recount an experience. SL 5: Add drawings or other visual displays to recounts of experiences. L 4: Determine or clarify the meaning of unknown or multiple-meaning words and phrases

Disciplinary Core Ideas		Crosscutting Concepts
PS1.A: Structure and properties of matter • Different kinds of matter exist and many of them can be either solid or liquid, depending on temperature. Matter can be described and classified by its observable properties. **(2-PS1-1)** • Different properties are suited to different purposes. **(2-PS1-2)**	**ETS1.A: Defining and delimiting engineering problems** • Before beginning to design a solution, it is important to clearly understand the problem. **(K-2-ETS1-1)** **ETS1.B: Developing possible solutions** • Designs can be conveyed through sketches, drawings, or physical models. These representations are useful in communicating ideas for a problem's solutions to other people. **(K-2-ETS1-2)** **ETS1.C: Optimizing the design solution** • Because there is always more than one possible solution to a problem, it is useful to compare and test designs. **(K-2-ETS1-3)**	Cause and effect Scale, proportion, and quantity Energy and matter
ESS1.C: The history of planet Earth • Some event happen very quickly; others occur very slowly over a time period much longer than one can observe. **(2-ESS1-1)** **ESS2.A: Earth materials and systems** • Wind and water can change the shape of the land. **(2-ESS2-1)** **ESS2.B: Plate tectonics and large-scale system interactions** • Maps show where things are located. One can map the shapes and kinds of land and water in any area. **(2-ESS2-2)** **ESS2.C: The roles of water in Earth's surface processes** • Water is found in the ocean, rivers, lakes, and ponds. Water exists as solid ice and in liquid form. **(2-ESS2-3)**	**ETS1.A: Defining and delimiting engineering problems** • Before beginning to design a solution, it is important to clearly understand the problem. **(K-2-ETS1-1)** **ETS1.B: Developing possible solutions** • Designs can be conveyed through sketches, drawings, or physical models. These representations are useful in communicating ideas for a problem's solutions to other people. **(K-2-ETS1-2)** **ETS1.C: Optimizing the design solution** • Because there is always more than one possible solution to a problem, it is useful to compare and test designs. **(K-2-ETS1-3)**	Cause and effect Scale, proportion, and quantity Stability and change

FOSS NEXT GENERATION K–8 SCOPE AND SEQUENCE

Grade	Physical Science	Earth Science	Life Science
6–8	Waves* Gravity and Kinetic Energy*	Planetary Science	Heredity and Adaptation* Human Systems Interactions*
	Chemical Interactions	Earth History	Populations and Ecosystems
	Electromagnetic Force* Variables and Design*	Weather and Water	Diversity of Life
5	Mixtures and Solutions	Earth and Sun	Living Systems
4	Energy	Soils, Rocks, and Landforms	Environments
3	Motion and Matter	Water and Climate	Structures of Life
2	Solids and Liquids	Pebbles, Sand, and Silt	Insects and Plants
1	Sound and Light	Air and Weather	Plants and Animals
K	Materials and Motion	Trees and Weather	Animals Two by Two

* Half-length course

Materials

Contents

INTRODUCTION

The Pebbles, Sand, and Silt kit contains

- *Teacher Toolkit: Pebbles, Sand, and Silt*
  - 1 *Investigations Guide: Pebbles, Sand, and Silt*
  - 1 *Teacher Resources: Pebbles, Sand, and Silt*
  - 1 *FOSS Science Resources: Pebbles, Sand, and Silt*
- *FOSS Science Resources: Pebbles, Sand, and Silt* (1 big book and class set of student books)
- Permanent equipment for one class of 32 students
- Consumable equipment for three classes of 32 students

FOSS modules use central materials distribution. You organize all the materials for an investigation on a single table called the materials station. As the investigation progresses, one member of each group gets materials as they are needed, and another returns the materials when the investigation is completed. You place items at the station—students do the rest.

Individual photos of each piece of FOSS equipment are available online for printing. For updates to information on materials used in this module and access to the Safety Data Sheets (SDS), go to www.FOSSweb.com. Links to replacement-part lists and customer service are also available on FOSSweb.

NOTE
To see how all of the materials in the module are set up and used, view the teacher preparation video on FOSSweb.

NOTE
Delta Education Customer Service can be reached at 1-800-258-1302.

KIT INVENTORY *List*

★ The student books, if included in your purchase, are shipped separately.

▶ NOTE
The teacher toolkit is shipped separately. However, there is space in drawer 1 to store your toolkit.

✪ These items might occasionally need replacement.

Drawer 1 of 3

Qty	Item	Equipment Condition
Print Materials		
1	*Teacher Toolkit: Pebbles, Sand, and Silt* (1 *Investigations Guide*, 1 *Teacher Resources*, and 1 *FOSS Science Resources: Pebbles, Sand, and Silt*)	
32	*FOSS Science Resources: Pebbles, Sand, and Silt*, student books*	
1	*FOSS Science Resources: Pebbles, Sand, and Silt*, big book	
1	Poster, *Natural Sources of Water*	
2	Posters, *FOSS Science Safety* and *FOSS Outdoor Safety*	
1	Poster set, *Conservation*, 4/set	
8	Sources of Water card sets, 18 cards/set	
Consumable Items		
1	Clay, potter's clay, box, 2.3 kg/box (5 lbs.)	
1	Clay, powdered, bag, 0.45 kg/bag (1 lb.)	
1	Potting soil, bag, 2 kg/bag (4.4 lbs.)	
5	Sand, clean, bags, 0.90 kg/bag (2 lbs.)	
4	Self-stick notes, pads, 100/pad	
50	Straws, jumbo	
Items for Investigation 1		
36	Construction paper sheets, black ✪	
72	Rocks, basalt (dark gray)	
20	Rocks, granite, pink, large	
72	Rocks, scoria (reddish)	
72	Rocks, tuff (yellowish)	
Items for Investigation 2		
36	Spoons, plastic	
Items for Investigation 3		
16	Aluminum mini loaf pans ✪	
100	Craft sticks ✪	
5	Sandpaper sheets, coarse, #50 ✪	
5	Sandpaper sheets, fine, #150 ✪	
5	Sandpaper sheets, medium, #80 ✪	

Drawer 2 of 3

		Equipment Condition
Shared Items		
8	Basins	
1	Bottle brush	
80	Containers, plastic, 1/4 L	
50	Cups, plastic, 250 mL (9 oz.)	
32	Hand lenses, 3-power	
2	Loupes/magnifying lenses	
1	Pitcher	
16	Spoons, metal	
8	Vial holders	
72	Vials, with caps, 12-dram	
75	Zip bags, 1 L	
10	Zip bags, 4 L	

▶ **NOTE**
This module includes access to FOSSweb, which includes the streaming videos and online activities used throughout the module.

Drawer 3 of 3

Shared Items (*continued*)		
1	Gravel, bag, 2.3 kg/bag (5 lbs.)	
1	Pebbles, large, bag, 2.3 kg/bag (5 lbs.)	
1	Pebbles, small, bag, 2 kg/bag (4.4 lbs.)	
70	Plates, paper, white ✪	
1	Sand, unwashed, with silt, bag, 2.3 kg/bag (5 lbs.) ✪	
16	Screen sets, 3 mesh sizes/set (large, medium, small mesh)	
1	Whisk broom and dustpan set	

✪ These items might occasionally need replacement.

> **NOTE**
> Throughout the *Investigations Guide*, we refer to materials not provided in the kit as "teacher-supplied." These materials are generally common or consumable items that schools and/or classrooms already have, such as rulers, paper towels, and computers. If your school/classroom does not have these items, they can be provided by teachers, schools, districts, or materials centers (if applicable). You can also borrow the items from other departments or classrooms, or request these items as community donations.

MATERIALS *Supplied by the Teacher*

Each part of each investigation has a Materials section that describes the materials required for that part. It lists materials needed for each student or group of students and for the class.

Be aware that you must supply some items. These are indicated with an asterisk (*) in the Materials list for each part of the investigation. Here is a summary list of those items by investigation.

For all investigations

- Chart paper and marking pen
- Drawing utensils (pencils, crayons, colored pencils, marking pens)
- Glue sticks
- Paper towels and/or sponges
- 1 Projection system or document camera (optional)
- Science notebooks (composition books)
- Water

For outdoor investigations

- 1 Bag for carrying materials
- 32 Clipboards
- Containers for water (plastic gallon jugs, 2 L soft-drink bottles)
- Pencils
- 1 Whistle or bell

Investigation 1: First Rocks

- Egg cartons (optional)
- 16 Pieces of white paper

Investigation 2: River Rocks

- 1 Large knife
- 32 Sheet protectors, clear (optional)
- Transparent tape

Investigation 3: Using Rocks

- • Clear acrylic spray (optional)
- 1 Brick
- 1 Bucket
- 1 Camera (optional)
- • Clay soil, 8 L
- 1 Container or jar with lid, 1 L
- • Containers for water
- 1 Box of cornstarch, 454 g (1 lb.)
- • White glue
- • Dry grass, straw, or weeds
- 1 Large knife
- • Newspaper
- 32 Paint brushes
- • Poster paints
- 1 Sandstone piece, landscaping (optional)
- 1 Saucepan
- 1 Scissors
- • Scratch paper
- 1 Mixing spoon
- 1 Plastic tablecloth or place mats (optional)
- • Tempera paints
- 1 Old hand towel (optional)
- 1 Trowel or shovel
- 16 Pieces of white paper
- • Yarn, 25 yards

Investigation 4: Soil and Water

- 1 Bucket
- • Scratch paper
- • Transparent tape
- • White paper (optional)

PREPARING *the Kit for Your Classroom*

Some preparation is required each time you use the kit. Doing these things before beginning the module will make daily setup quicker and easier.

1. Inventory materials

Before using a kit, conduct a quick inventory of all items in the kit. You can use the list provided in this chapter to keep track of any items that are missing or in need of replacement. Information on ordering replacement items can be found at the end of this chapter. The kit contains enough consumables for at least three classes of 32 students.

Basalt

2. Inventory Sources of Water cards

One set of Sources of Water cards contains 18 different cards (two cards each of these nine sources: coral reefs, glaciers and ice, mangrove forests, ocean and seas, ponds and lakes, rocky coasts, salt marshes, sandy beaches, and streams and rivers). Inventory each of the eight sets of cards.

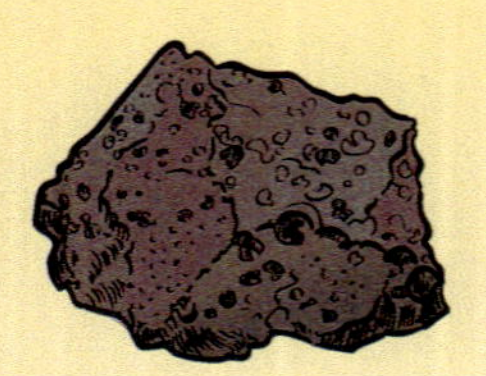

Scoria

3. Check permanent earth materials

The following rocks included in the kit are permanent equipment that should serve countless classes over the years:

Basalt	72 pieces (dark gray rock)
Scoria	72 pieces (reddish or gray lava rock)
Tuff	72 pieces (yellowish or light-colored rock)
Pink granite	20 large pieces

Tuff

About 300 **large pebbles** are used in Investigation 1, Part 3. Half of these pebbles are used to make the river rock mixture in Investigation 2 but are not consumed.

Pink granite

4. Check the consumable earth materials

A number of items in the kit are consumable. All these materials can be purchased from local sources, and replacement packages for this module are available from Delta Education.

Gravel is used to make the river rock mixture in Investigation 2 but is not consumed. An insignificant amount of the gravel might be glued to students' records of rock sizes in Investigation 2. In Investigation 4, a cup of gravel is used to make soil, and this should be considered consumable, as it is very tedious to reclaim.

Small pebbles are used to make the river rock mixture in Investigation 2 but are not consumed. A small number of pebbles are glued to students' permanent records of rock sizes in Investigation 2 and this should be considered consumable, as it is very tedious to reclaim. In Investigation 4, a cup of small pebbles is used per class to make the soil.

Unwashed sand, which contains some silt and clay, is used in Investigations 2 and 4. Some of this sand will be consumed as the module progresses.

Clean sand (washed sand) is used for sand sculptures and is in the kit in a modest supply. Clean sand is readily available at toy, pet, and garden supply stores. You will need about 1.5 liters (2 kilograms or 4 lb.) for Investigation 3.

Moist **potter's clay** is used in Investigation 2. It is possible to recycle some of this clay.

A small bag of **dry powdered clay** is included for replenishing the silt in the sand used in Investigation 2. In time it will be used up.

5. Acquire more earth materials

Most of the earth materials used in this module can be purchased from local home building and garden supply stores. Search the Internet using "sand and gravel" or "rock." These stores would make great field-trip destinations for the class. Clay can be purchased at art and craft stores.

6. Plan for cleanup

This module uses a lot of containers and tools that must be cleaned, dried, inventoried, and stored right after each session if they are to stay in top condition. This responsibility should be assumed by students. Assign one collaborative group to do this after each session. Plan where this cleanup will take place in the classroom (if you have a sink) or elsewhere in the school if a sink is not available in your room.

All the plastic items in the kits are meant to be cleaned and reused.

Also plan for the disposal of liquids. A collection bucket works well when there is no sink. Any water that contains earth materials (soil, sand, clay) should not be poured down sink drains.

7. Print or photocopy notebook sheets

You will need to print or make copies of science notebook sheets before each investigation. See Getting Ready for Investigation 1, Part 1, for ways to organize the notebook sheets for this module. If you use a projection system, you can download electronic copies of the sheets from FOSSweb.

EL NOTE

You may want to print out the FOSS equipment photo cards (from FOSSweb) to add to your word wall to help students with vocabulary.

8. Plan for word wall and pocket charts

As the module progresses, you will add new vocabulary words to a word wall or pocket chart and model writing and responding to focus questions. See Investigation 1, Part 1, Getting Ready, Step 8, for suggestions about how to do this in your classroom.

9. Plan for safety issues indoors and outdoors

Two safety posters are included in the kit—*Science Safety* and *Outdoor Safety*. You should review the guidelines with students and post the posters in the room as a reminder. Getting Ready for Investigation 1, Part 1, offers suggestions for this discussion. Emphasize that materials do not go in mouths, ears, noses, or eyes. Encourage responsible actions toward other students.

Also be aware of any allergies that students in your class might have. Students will be working with sand and clay that can result in dust particles. Caution students not to rub their eyes with sandy hands.

Use the four *Conservation* posters to discuss the importance of conserving natural resources.

NOTE
The ***Letter to Family*** and ***Home/School Connections*** are available electronically on FOSSweb.

10. Plan for letter home and home/school connections

Teacher master 1, *Letter to Family*, is a letter you can use to inform families about this module. The letter states the goals of the module and suggests some home experiences that can contribute to students' learning.

There is a home/school connection for most investigations. Check the last page of each investigation for details, and plan when to print or make copies and send them home with students.

11. Gather books from library

Check your local library for books related to this module. Visit FOSSweb for a list of appropriate trade books that relate to this module.

12. Check FOSSweb for resources

Go to FOSSweb, register as a FOSS teacher, and review the print and digital resources available for this module, including the eGuide, eBook, Resources by Investigation, and *Teacher Resources*, including the grade-level Planning Guide. Be sure to check FOSSweb often for updates and new resources.

CARE, *Reuse, and Recycling*

When you finish teaching the module, inventory the kit carefully. Note the items that were used up, lost, or broken, and immediately arrange to replace the items. Use a photocopy of the Kit Inventory List and put your marks in the "Equipment Condition" column. Refill packages and replacement parts are available for FOSS by calling Delta Education at 1-800-258-1302 or by using the online replacement-part catalog (www.DeltaEducation.com/RefillCenter).

Standard refill packages of consumable items are available from Delta Education. A refill package for a module includes sufficient quantities of all consumable materials (except those provided by the teacher) to use the kit with three classes of 32 students.

Here are a few tips on storing the equipment after use.

- Make sure items are clean and dry before storing them.
- Make sure the posters and print materials are flat on the bottom of the box.
- Inventory and bag up the water source cards.
- Sort and inventory all items and secure them in plastic bags.
- Keep the sets of small rocks (2 basalt, 2 scoria, and 2 tuff) in individual zip bags.
- Double-bag the river rock mixture from Investigation 2 in two large zip bags. Label them "River Rock."
- Use the bottle brush and hot water to clean the vials and other containers. Be sure they are completely dry before storing them in the kit.
- Check that the bags of sand and soil are tightly sealed, so they won't leak.
- Recycle clay that can be used again.

The items in the kit have been selected for their ease of use and durability. Small items should be inventoried (a good job for students under your supervision) and put into zip bags for storage. Any items that are no longer useful for science should be properly recycled.

Technology

You are not logged in: Login or Register
Just Looking? Visit as a Guest
FOSS
Home
About FOSS
Teaching Tools
Materials Management
Connecting with FOSS
Search...
Welcome to FOSSweb
The Full Option Science System™ website
FOSSweb is the official website of the active-learning science program, FOSS. Explore resources for educators and engaging activities for students and families.
To get started, log in or visit as a guest. Students may visit as a guest, or use the class login if you have one.
Teacher Login
Class Login
Or, Visit as a Guest
Username or email
Password
Remember Me
Teachers only: I've forgotten my password.
Log In
Don't have an account yet? Get started now.
What is FOSS?
FOSS is a research-based science curriculum for grades K-8 developed at the Lawrence Hall of Science, University of California, Berkeley.
Find out more...

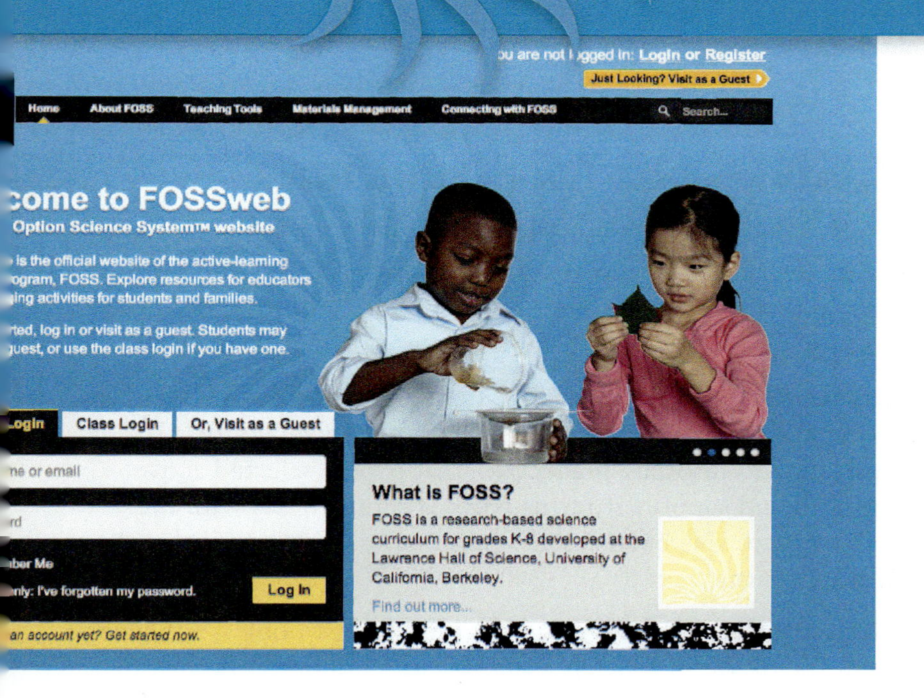

Contents

INTRODUCTION

Technology is an integral part of the teaching and learning with FOSS Next Generation. FOSSweb is the Internet access to FOSS digital resources. FOSSweb gives students the opportunity to interact with simulations, virtual investigations, tutorials, images, and text—activities that enhance understanding of core ideas. It provides support for teachers, administrators, and families who are actively involved in implementing FOSS.

Different types of online activities are incorporated into investigations where appropriate. Each activity is marked with the technology icon in the *Investigations Guide*. You will sometimes show videos to the class. At other times, individuals or small groups of students will work online to review concepts or reinforce their understanding.

To use these digital resources, you should have at least one computer with Internet access that can be displayed to the class by an LCD projector with an interactive whiteboard or a large screen. Access to enough devices for students to work in small groups or one-on-one is recommended for other parts.

All FOSS online activities are available at www.FOSSweb.com for teachers, students, and families. We recommend you access FOSSweb well before starting the module to set up your teacher-user account and to become familiar with the resources.

NOTE
To get the most current information, download the latest Technology chapter on FOSSweb.

TECHNOLOGY *for Students*

FOSS is committed to providing a rich, accessible technology experience for all FOSS students. Students access FOSSweb using a class login that you set up. Here are brief descriptions of selected resources for students on FOSSweb.

Online activities. The online simulations and activities are designed to support students' learning at all grades. They include virtual investigations and tutorials, grades 3–5, that review selected active investigations and support students who have difficulties with the materials or who have been absent. Summaries of some of the online activities are on the next page.

FOSS Science Resources—*eBooks.* The student book is available as an audio book on FOSSweb, accessible at school or at home. In addition, as premium content, *FOSS Science Resources* is available as an eBook on computer or tablet, either as a read-only PDF or in an interactive format that allows text to be read and provides points of interactivity. The eBook can also be projected for guided reading with the whole class.

Media library. A variety of media enhances students' learning and provides them with opportunities to obtain, evaluate, and communicate information. FOSS has reviewed print books and digital resources that are appropriate for students and prepared a list of these resources with links to content websites. There is also a list of regional resources for virtual and actual field trips for students to use in gathering information for projects, and a database of science and engineering careers. Other resources include vocabulary lists to promote use of academic language.

Home/school connections. Each module includes a letter to families, providing an overview of the goals and objectives of the module. There is also a Module Summary available for families to download. Most investigations have a home/school science activity that connects the classroom experiences with students' lives outside of school. These connections are available as PDFs on FOSSweb.

Class pages. Teachers with a FOSSweb account can easily set up class pages with notes and assignments for each class. Students and families can then access this class information online, using the teacher-assigned class login.

NOTE

The following student-facing resources are available in Spanish on FOSSweb using a teacher's class page.

- Vocabulary
- Equipment photo cards
- eBooks
- Select streaming videos
- Home/school connections
- Audio books

Pebbles, Sand, and Silt Online Activities

Here is a sampling of the online activities used in the **Pebbles, Sand, and Silt Module** investigations.

Investigation 1, Part 5: Sorting Activities

- **"Rock Sorting"**
  Students are presented with a virtual rock collection. They determine a property for a group and then move the rocks that exhibit that property into the bucket. Students can reset and regroup the groups based on new properties.

Investigation 1, Part 5: Sorting Activities

- **"Property Chain"**
  Pairs of students work together to build a chain of rocks with similar properties! Player 1 places a rock in the path. Player 2 picks a rock that shares one property of Player 1's rock and names the property they were thinking of when they placed the rock in the chain. Students continue taking turns building a chain of rocks that share a property with the rock before it. The pair of students see how long of a chain they can build.

Investigation 3, Part 1: Rocks in Use

- **"Find Earth Materials"**
  Students search for products and objects made of Earth materials in two outdoor scenes, one in a park and one along a sidewalk scene. At any time, students can use the reference to see and read about each rock size: clay, silt, sand, gravel, pebbles, cobbles, and boulders.

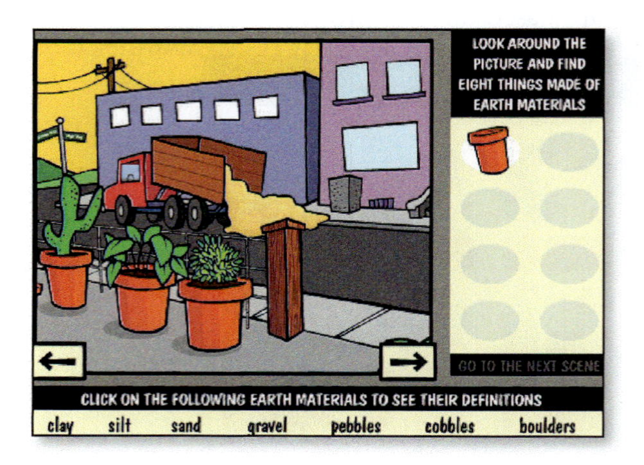

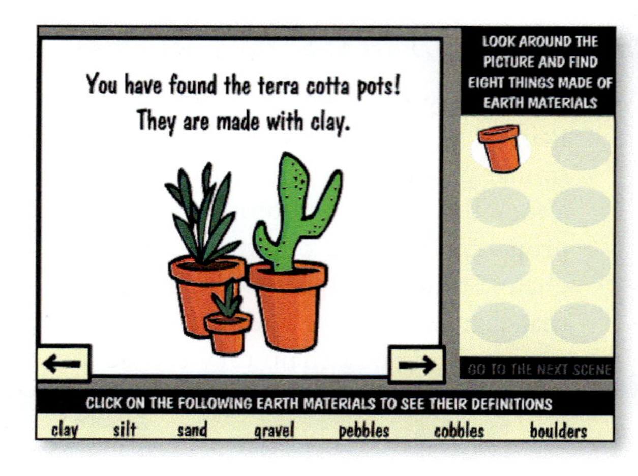

TECHNOLOGY *for Teachers*

The teacher side of FOSSweb provides access to all the student resources plus those designed for teaching FOSS. By creating a FOSSweb user account and activating your modules, you can personalize FOSSweb for easy access to your instructional materials. You can also set up a class login for students and their families.

Creating a FOSSweb Teacher Account

Setting up an account. Set up a teacher account on FOSSweb before you begin teaching a module. Go to FOSSweb and register for an account with your school e-mail address. Complete registration instructions are available online. If you have a problem, go to the Connecting with FOSS pull-down menu, and look at Technical Help and Access Codes. You can also access online tutorials for getting started with FOSSweb at www.FOSSweb.com/fossweb-walkthrough-videos.

Entering your access code. Once your account is set up, go to FOSSweb and log in. To gain access to all the teacher resources for your module, you will need to enter your access code. Your access code should be printed on the inside cover of your *Investigations Guide*. If you cannot find your FOSSweb access code, contact your school administrator, your district science coordinator, or the purchasing agent for your school or district.

Familiarize yourself with the layout of the site and the additional resources available when you log in to your account. From the module detail page, you will be able to access teacher masters, science notebook masters, assessment masters, the FOSSmap online assessment component, and other digital resources not available to "guests."

Explore Resources by Investigation, as this will help you plan. This page makes it simple to select the investigation you are teaching, and view all the digital resources organized by part. Resources by Investigation provides immediate access to the streaming videos, online activities, science notebook masters, teacher masters, and other digital resources for each investigation part.

Setting up class pages and student accounts. To enable your students to log in to FOSSweb to see class assignments and student-facing digital resources, set up a class page and generate a username and password for the class. To do this, log in to FOSSweb and go to your teacher page. Under "My Class Pages," follow the instructions to create a new class page and to leave notes for students. Note: student access to the student eBook from your class page requires premium content.

NOTE

For more information about FOSS premium content, including pricing and ordering, contact your local Delta sales representative by visiting www.DeltaEducation.com or by calling 1-800-258-1302.

Support for Teaching FOSS

FOSSweb is designed to support teachers using FOSS. FOSSweb is your portal to instructional tools to make teaching efficient and effective. Here are some of the tools available to teachers.

- **Grade-level Planning Guide.** The Planning Guide provides strategies for three-dimensional teaching and learning.
- **Resources by Investigation.** The Resources by Investigation organizes in one place all the print and online instructional materials you need for each part of each investigation.
- **Investigations eGuide.** The eGuide is the complete *Investigations Guide* component of the *Teacher Toolkit*, in an electronic web-based format for computers or tablets. If your district rotates modules among several teachers, this option allows all teachers easy access to *Investigations Guide* at all times.
- **Teacher preparation videos.** Videos present information to help you prepare for a module, including detailed investigation information, equipment setup and use, safety, and what students do and learn in each part of the investigation.
- **Interactive whiteboard resources for grades K–5.** You can use these interactive files with or without an interactive whiteboard to facilitate each part of each investigation. You'll need to download the appropriate software to access the files. Links for software downloads are on FOSSweb.
- **Focus questions.** The focus questions address the phenomenon for each part of each investigation, and are formatted for classroom projection and for printing, so that students can glue each focus question into their science notebooks.
- **Module updates.** Important updates cover teacher materials, student equipment, and safety considerations.
- **Module teaching notes.** These notes include teaching suggestions and enhancements to the module, sent in by experienced FOSS users.
- **Home/school connections.** These masters include an introductory letter home (with ideas to reinforce the concepts being taught) and the home/school connection sheets.
- **State and regional resources.** Listings of resources for your geographic region are provided for virtual and actual field trips and for students to use as individual or class projects.
- **Access to FOSS developers.** Through FOSSweb, teachers have a connection to the FOSS developers and expert FOSS teachers.

NOTE
There are two versions of the eGuide, a PDF-based eGuide that mimics the hard copy guide, and an HTML interactive eGuide that allows you to write instructional notes and to interface with online resources from the guide.

NOTE
The following resources are available on FOSSweb in Spanish.

Teacher-facing resources:
- Notebook masters
- Teacher masters
- Assessment masters
- Focus questions
- Interactive whiteboard files

Student-facing resources:
- Vocabulary
- Equipment photo cards
- eBooks
- Select streaming videos
- Home/school connections
- Audio books

Technology for Differentiated Instruction

Some resources are for differentiated instruction. They can be used by students at home or by you as part of classroom instruction.

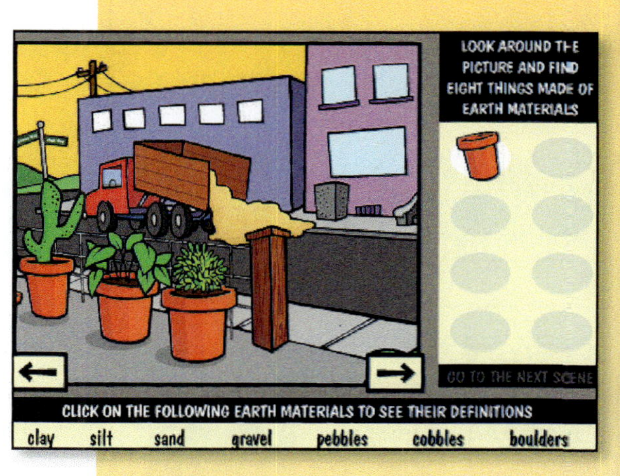

- **Online activities.** The online simulations and activities described earlier in this chapter are designed to support student learning and are often used during instruction. They include virtual investigations and student tutorials for grades 3–5 that you can use to support students who have difficulties with the materials or who have been absent. Tutorials require students to record data and answer concluding questions in their notebooks. In some cases, the notebook sheet used in the classroom investigation is also used for the virtual investigation.
- **Vocabulary.** The online word list has science-related vocabulary and definitions used in the module (in both English and Spanish).
- **Equipment photo cards.** Equipment cards provide labeled photos of equipment that students use in the investigations. Cards can be printed and posted on the word wall as part of instruction.
- **Student eBooks.** Student access to audio-only *FOSS Science Resource* books requires basic access. With premium content, students can access the books from any Internet-enabled device. The eBooks are available in PDF and interactive versions. The PDF version mimics the hard copy book. The interactive eBook reads the text to students—highlighting the text as it is read—and provides students with video clips and online activities.
- **Streaming videos used for extensions.** Some videos are part of the instruction in the investigation and are in Resources by Investigation for each part. Those videos also appear again in the digital resources under "Streaming Videos" along with other videos that extend concepts presented in a module.
- **Recommended books, websites, and careers database.** FOSS-recommended books, websites, and a Science and Engineering Careers Database that introduces students to a variety of career options and diversity of individuals engaged in those careers are provided.
- **Regional resources.** This list provides local resources that can be used to enhance instruction. The list includes website links and PDF documents from local sources.

NOTE
The eBook is premium content for students.

Support for Classroom Materials Management

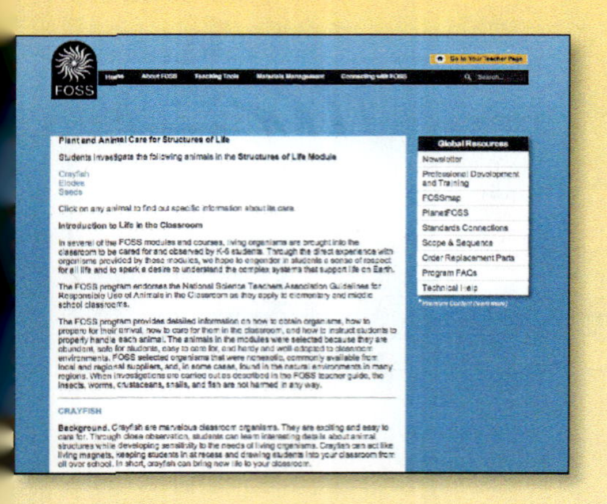

- **Materials chapter.** A PDF of the Materials chapter in *Investigations Guide* is available to help you prepare for teaching. A list, organized by drawer, shows the materials included in the FOSS kit for a given module. You can print and use this list for inventory and to monitor equipment condition.
- **Safety Data Sheets (SDS).** A link takes you to the latest safety sheets, with information from materials manufacturers on the safe handling and disposal of materials.
- **Plant and animal care.** This section includes information on caring for organisms used in the investigations.

Professional Learning Connections

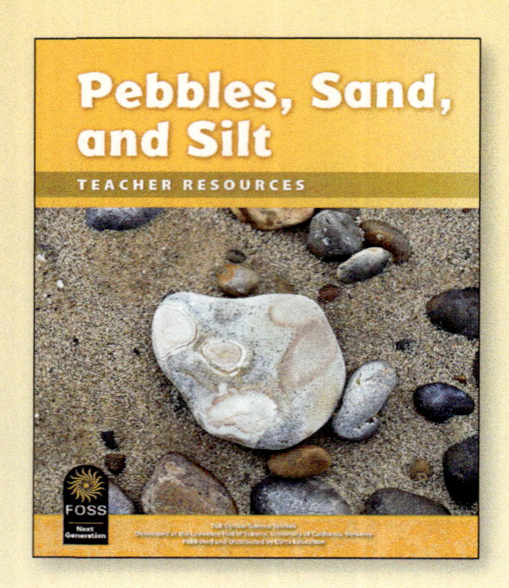

FOSSweb provides PDF files of professional development chapters, mostly from *Teacher Resources*, that explain how to integrate instruction to improve learning. Some of them are

- Sense-Making Discussions for Three-Dimensional Learning
- Science-Centered Language Development
- FOSS and Common Core English Language Arts and Math
- Access and Equity
- Taking FOSS Outdoors

REQUIREMENTS *for Accessing FOSSweb*

FOSSweb Technical Requirements

To use FOSSweb, your computer must meet minimum system requirements and have a compatible browser and recent versions of Flash Player, QuickTime, and Adobe Reader. Many online activities have been updated to an HTML5 version compatible with all devices. (Those designated with "Flash" after the title require Flash Player.) The system requirements are subject to change. It is strongly recommended that you visit FOSSweb to review the most recent minimum system requirements and any plug-in requirements. There, you can access the "Tech Specs and Info" page to confirm that your browser has the minimum requirements to support the online activities.

Preparing your browser. FOSSweb requires a supported browser for Windows or Mac OS with a current version of the Flash Player plug-in, the QuickTime plug-in, and Adobe Reader or an equivalent PDF reader program. You may need administrator privileges on your computer in order to install the required programs and/or help from your school's technology coordinator.

By accessing the "Tech Specs and Info" page on FOSSweb, you can check compatibility for each computer you will use to access FOSSweb, including your classroom computer, computers in a school computer lab, and a home computer. The information on FOSSweb contains the most up-to-date technical requirements for all devices, including tablets and mobile devices.

Support for plug-ins and reader. Flash Player and Adobe Reader are available on www.adobe.com as free downloads. QuickTime is available for free from www.apple.com. FOSS does not support these programs. Please go to the program's website for troubleshooting information.

NOTE
It is strongly recommended that you visit FOSSweb to review the most recent minimum system requirements.

Other FOSSweb Considerations

Firewall or proxy settings. If your school has a firewall or proxy server, contact your IT administrator to add explicit exceptions in your proxy server and firewall for FOSSweb Akamai video servers. For more specific information on servers for firewalls, refer to "Tech Specs and Info" on FOSSweb.

Classroom technology setup. FOSS has a number of digital resources and makes every effort to accommodate users with different levels of access to technology. The digital resources can be used in a variety of ways and can be adapted to a number of classroom setups.

Teachers with classroom computers and an LCD projector, interactive whiteboard, or a large screen will be able to show online materials to the class. If you have access to a computer lab, or enough computers in your classroom for students to work in small groups, you can set up time for students to use the FOSSweb digital resources during the school day. Teachers who have access to only a single computer will find a variety of resources on FOSSweb that can be used to assist with teacher preparation and materials management.

Teachers who have tablets available for student use and have premium content can download the FOSS eBook app onto devices for easy student access to the FOSS eBooks. Instructions for downloading the app can be found on FOSSweb on the Module Detail Page for any module. You'll find them under the Digital-Only Resources section and then under the tab for Student eBooks.

NOTE
FOSSweb activities are designed for a minimum screen size of 1024 × 768. It is recommended that you adjust your screen resolution to 1024 × 768 or higher.

Displaying online content. Throughout each module, you may occasionally want to project online components for instruction through your computer. To do this, you will need a computer with Internet access and either an LCD projector and a large screen, an interactive whiteboard, or a document camera arranged for the class to see.

You might want to display the notebook and teacher masters to the class. In Resources by Investigation, you'll have the option of downloading the masters to project or to copy. Choose "for Display" if you plan on projecting to the class. These masters are optimized for a projection system and allow text entry directly onto the sheet from the computer. The "for Print" versions are sized to minimize paper use when photocopying for the class. In Resources by Investigation the print versions of the masters are typically unlabeled.

TROUBLESHOOTING *and Technical Support*

If you experience trouble with FOSSweb, you can troubleshoot in a variety of ways.

1. Test your browser to make sure you have the correct plug-in and browser versions. Even if you have the necessary plug-ins installed on your computer, they may not be recent enough to run FOSSweb correctly. Go to FOSSweb, and select the "Tech Specs and Info" page to review the most recent system requirements and check your browser.
2. Check the FAQs on FOSSweb for additional information that may help resolve the problem.
3. Empty the cache from your browser and/or quit and relaunch.
4. Restart your computer, and make sure all computer hardware turns on and is connected correctly.

If you are still experiencing problems after taking these steps, send FOSS Technical Support an e-mail to support@FOSSweb.com. In addition to describing the problem you are experiencing, include the following information about your computer: Mac or PC, operating system, browser name and version, plug-in names and versions. This will help us troubleshoot the problem.

NOTE
The FOSS digital resources are available online on FOSSweb. You can always access the most up-to-date technology information, including help and troubleshooting, on FOSSweb.

Where to Get Help

For further questions about FOSSweb, please don't hesitate to contact our technical support team.

Account questions/help logging in

School Specialty Online Support
techsupport.science@schoolspecialty.com
loginhelp@schoolspecialty.com

Phone: 1-800-513-2465, 8:30 a.m.–6:00 p.m. ET

General FOSSweb technical questions

FOSSweb Tech Support
support@fossweb.com

Investigation 1: First Rocks

INVESTIGATION 1 — *First Rocks*

Guiding question for phenomenon: ***What are properties of rocks and how do they change?***

PURPOSE

Students investigate several different rocks. In the process, they are introduced to the phenomenon that rocks are not all the same. After rubbing two samples together, students find that rock is hard but also susceptible to weathering.

Content

- Rocks can be described by their properties.
- Smaller rocks (sand) result from the breaking (weathering) of larger rocks.
- Rocks are the solid material of Earth. Rocks are composed of minerals.
- Some rocks (such as tuff, scoria, and basalt) are formed from lava and other materials produced by erupting volcanoes. Volcanoes are mountains built by melted rock that flow out of weak areas in Earth's crust.

Practices

- Use tools and water to compare properties of rocks.
- Compare and sort rocks in different ways, using two or more physical properties.

Science and Engineering Practices

- Asking questions
- Planning and carrying out investigations
- Analyzing and interpreting data
- Constructing explanations
- Engaging in argument from evidence
- Obtaining, evaluating, and communicating information

Disciplinary Core Ideas

ESS1: What is the universe, and what is Earth's place in it?
ESS1.C: The history of planet Earth
PS1: How can one explain the structure, properties, and interactions of matter?
PS1.A: Structure and properties of matter

Crosscutting Concepts

- Patterns
- Cause and effect
- Stability and change

INVESTIGATION 1 — *First Rocks*

	Investigation Summary	Time	Focus Question for Phenomenon, Practices
PART 1	**Three Rocks** Students investigate and sort a set of six rocks. They gather information about the rocks by observing and comparing, then rub them together to simulate weathering.	**Active Inv.** 1 Session*	**What happens when rocks rub together?** **Practices** Planning and carrying out investigations Analyzing and interpreting data Constructing explanations
PART 2	**Washing Three Rocks** Students wash their rocks to see how they change when they are wet, and to see what happens to the wash water. Students are introduced to the names of these volcanic rocks (tuff, scoria, basalt) and view a video on volcanoes to find out how they formed.	**Active Inv.** 1 Session	**What happens when rocks are placed in water?** **Practices** Planning and carrying out investigations Analyzing and interpreting data Constructing explanations Obtaining, evaluating, and communicating information
PART 3	**First Sorting** Students are introduced to river rocks, describe their properties, and compare and sort them into groups based on one property at a time.	**Active Inv.** 1 Session	**How are river rocks the same?** **Practices** Planning and carrying out investigations Analyzing and interpreting data Constructing explanations Engaging in argument from evidence
PART 4	**Start a Rock Collection** Students take a field trip to collect and observe schoolyard rocks. They describe the properties of the various rocks.	**Active Inv.** 2 Sessions **Reading** 1 Session	**What are the properties of schoolyard rocks?** **Practices** Asking questions Analyzing and interpreting data Obtaining, evaluating, and communicating information
PART 5	**Sorting Activities** Students use sorting mats to compare and sort the river rocks.	**Active Inv.** 1 Session **Reading** 1 Session **Assessment** 1 Session	**How many ways can rocks be sorted?** **Practices** Analyzing and interpreting data Obtaining, evaluating, and communicating information

* A class session is 45–50 minutes.

Content Related to DCIs	Writing/Reading	Assessment
• Rocks can be described by their properties. • Smaller rocks (sand) result from the breaking (weathering) of larger rocks.	**Science Notebook Entry** *Rubbing Rocks*	**Embedded Assessment** Science notebook entry
• Rocks can be described by their properties. • When rocks are washed in water, the colors or sparkling qualities are enhanced. • Some rocks (such as tuff, scoria, and basalt) are formed from lava and other materials produced by erupting volcanoes. • Volcanoes are mountains built by melted rock that flow out of weak areas in Earth's crust.	**Science Notebook Entry** *Rocks in Water* **Video** *All about Volcanoes*	**Embedded Assessment** Science notebook entry
• Rocks can be sorted by their properties. • When rocks are washed in water, the colors or sparkling qualities are enhanced.	**Science Notebook Entry** *Rock Sorting*	**Embedded Assessment** Performance assessment
• Rocks are all around us. • Rocks are the solid material of Earth.	**Science Notebook Entry** *Rock Record* **Science Resources Book** "Exploring Rocks"	**Embedded Assessment** Science notebook entry
• Rocks can be described by their properties. • Rocks are composed of minerals.	**Science Resources Book** "Colorful Rocks" **Online Activities** "Rock Sorting" "Property Chain"	**Benchmark Assessment** *Investigation 1 I-Check* **NGSS Performance Expectations addressed in this investigation** 2-ESS1-1 2-PS1-1

TEACHING NOTE

Refer to the grade-level Planning Guide chapter in Teacher Resources *for a summary explanation of the phenomena students investigate in this module using a three-dimensional learning approach.*

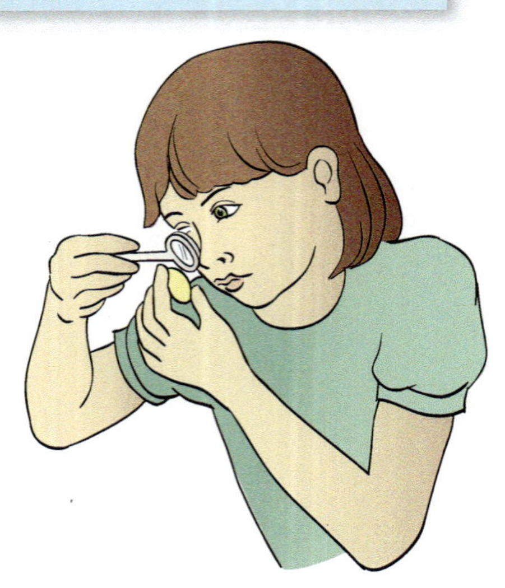

BACKGROUND *for the Teacher*

The anchor phenomena for this module is earth material that covers our planet's surface. One of those earth materials is rock. Many children at one time or another start rock collections. Some collections never extend beyond the pocket of a pair of jeans or a modest heap in the corner of a treasure box, but others take prominent places on windowsills or bookshelves at home. "The quiet company of rocks" may capture primary students' imaginations and spark an interest that endures for a lifetime.

What is it about rocks that makes them so collectible? They're durable, sometimes **shiny**, and occasionally sparkly. They come in different **sizes**, **shapes**, and **colors**, and they can look like jewels, especially when placed in water. You don't have to feed them, and they don't rot, dry out, or die if you forget to care for them.

Earth materials are any of the solid, liquid, or gaseous nonliving materials that make up Earth. **Rocks** are natural, solid earth materials made of **minerals**. Although there are thousands of minerals and even more types of rock, there are only about 20 common minerals and perhaps three times that many kinds of rock that could be considered common.

Geologists, the scientists who study Earth, are especially intrigued by the differences in rock composition. When they come across a rock that is new to them, they observe the **properties** of the rock and try to determine its story—its composition and history. Properties such as color, hardness, and luster of the minerals in a rock, and the possible presence of fossils, help a geologist determine a rock's composition. Knowing a rock's composition can help a geologist figure out how, where, and when the rock formed. Sometimes the properties of one particular kind of rock can be so unusual that a geologist can figure out the exact mountain or cliff that the rock came from, even if that mountain is hundreds of kilometers away.

Observing rocks and beginning to sort them into **groups** are the initial steps students take in their role as geologists. Observing is a fundamental process in geology. Through observation, students gather firsthand information (**data**) about rocks. They use these observations to make comparisons and to **sort** the rocks into groups having similar properties.

What Happens When Rocks Rub Together?

Students explore several kinds of rocks in this investigation. The first set of six rocks has three types of volcanic rock, chosen because of their obvious differences in color and **texture**. These include **basalt**, a dark, dense, fine-grained rock that began as molten lava extruded from a

volcano; **tuff**, a light-colored, soft rock composed of tiny fragments of volcanic ash; and **scoria**, a reddish porous rock sometimes with **sharp** edges, formed from the froth on the top of flowing lava. When tuff rocks are rubbed together, soft powder or dust comes off. When scoria rocks are rubbed together, the bits of rock that result might be bigger and **rougher**, looking more like **sand**. Basalt is a very hard rock, and very little comes off when pieces of basalt are rubbed together. When harder rocks are rubbed against softer rocks, the softer rocks break or weather. **Weathering** is the word used to describe the breakdown of rock into smaller pieces. Weathering is a slow Earth changing process in comparison to volcanic activity that causes rapid changes to Earth's surface.

Basalt

Tuff

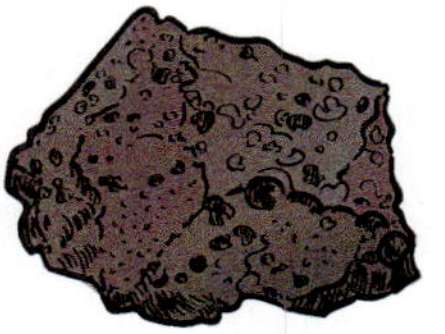

Scoria

What Happens When Rocks Are Placed in Water?

When rocks are wet, **dull** becomes shiny, and **patterns** in the surface emerge. Color may change, stripes may become more visible, and air pockets are revealed when **bubbles** escape. Putting rocks in water also washes away the weathered rock bits, and the water may become turbid.

"Even experienced geologists can surprise themselves when they observe the same rock a second time."

How Are River Rocks the Same?

The selection of river rocks provided in this module can include a wide variety of rocks, depending on where the rocks were gathered. You can usually find samples of the harder types of rocks, such as **granite** and quartzite, in this collection. Students will consider a small assortment of these river rocks. Based on size, color, texture, shape, or pattern, they will select one property and separate a group that is the same—perhaps finding rocks that are similarly **flat**, **pointed**, **round**, or **smooth**—it will depend on the rocks in your kit and your students.

What Are the Properties of Schoolyard Rocks?

Students will look at rocks in the schoolyard through a different lens in this module. It is important to start the class rock collection with the schoolyard rocks and add to it throughout the module. Students can act as geologists by observing and describing properties and sorting the schoolyard rocks into groups.

How Many Ways Can Rocks Be Sorted?

Students often sort rocks by color first. With further challenges they start to notice subtle differences between rocks and begin sorting by properties such as shape, texture, weight, luster, and pattern. Even experienced geologists can surprise themselves when they observe the same rock a second time. Unexpected rock properties they hadn't noticed before jump out—rock-shock—ready to be recognized and appreciated.

TEACHING CHILDREN *about First Rocks*

NGSS Foundation Box for DCI

ESS1.C: The history of planet Earth

- Some events happen very quickly; others occur very slowly over a time period much longer than one can observe. (2-ESS1-1)

PS1.A: Structure and properties of matter

- Different kinds of matter exist and many of them can be either solid or liquid, depending on temperature. Matter can be described and classified by its observable properties. (2-PS1-1)

Developing Disciplinary Core Ideas (DCI)

Richness of understanding comes from depth of experience. FOSS investigations guide students through a number of simple experiences, any one of which in isolation may appear overly simple or trivial. In fact this conclusion might be valid, if only one or two of the experiences are provided for students. But the power of FOSS is derived from the accumulation of experiences, acquired over time, that help students construct a deeper understanding of the important ideas of science that are embedded in the program.

Primary students can recognize a rock. It is a part of the childhood experience regardless of the culture and environment in which children live. Rocks are just that common and that important in the human experience.

Students at this age are ready to refine and expand their understanding of rocks. They should be exposed to a number of different kinds of rocks, so that a concept of "rockness" begins to emerge. When rock takes form as a *concept,* students will understand that rock is a material that can take a multitude of forms, rather than simply a collection of individual objects.

One key to helping primary students engage in real science is to ensure that the activities are geared to their level of interest and understanding. Primary students will be delighted to interact with a handful of rocks for an extended period of time, banging and rubbing them together, rolling them around, putting them in a row, shaking them in a cup, comparing them to their neighbor's, washing them, drying them, finding little marks or patterns, and who knows what else. It is through actions and transformations that children form a lasting understanding of their world.

Primary students are equipped to observe objects carefully, compare and communicate their discoveries, and sort objects to understand how things go together into groups. When students are challenged to interact with materials using these cognitive processes, learning will be enjoyable.

Primary students are still developing the coordination and dexterity needed to handle materials efficiently. Even when they are careful, they will spill rocks and water from time to time, and the floor will doubtless get a bit gritty. In short, students are likely to make a mess. However, you shouldn't let this deter you from providing these kinds of valuable experiences for students. The mess can be cleaned up as part of the overall learning experience. A whisk broom and dustpan will become part of the classroom equipment, and everyone should be adept at using

them. Cleaning up can contribute to a complete understanding of the properties of the materials.

The notion of the time involved to change the surface of Earth is a difficult one for young students to understand but it is central to this module. They are introduced to this concept of time by having firsthand experience with weathering of rocks to produce just a little bit of sand and comparing that to the long periods of time that it takes rocks to weather mountains in nature to produce the sand we find on beaches and in the soil. Students also experience rapid changes to Earth's surface virtually through videos of volcanic eruptions and compare that to the long process of weathering of rock.

The activities and media students engage with in this first investigation contribute to the disciplinary core ideas **ESS1.C, The history of planet Earth** and **PS1.A, Structure and properties of matter.**

Engaging in Science and Engineering Practices (SEP)

Engaging in the practices of science helps students understand how scientific knowledge develops; such direct involvement gives them an appreciation of the wide range of approaches that are used to investigate, model, and explain the world. Engaging in the practices of engineering likewise helps students understand the work of engineers, as well as the links between engineering and science. Participation in these practices also helps students form an understanding of the crosscutting concepts and disciplinary ideas of science and engineering; moreover, it makes students' knowledge more meaningful and embeds it more deeply into their worldview. (National Research Council, *A Framework for K-12 Science Education*, 2012, page 42)

In this investigation, students engage in these practices.

- **Asking questions** about rocks and rock properties that are found.
- **Planning and carrying out investigations** of volcanic rocks and river rocks to make comparisons and to understand how they were formed.
- **Analyzing and interpreting data** by describing observations of the different kinds and sizes of rocks and making rock records using words and pictures.
- **Constructing explanations** by making firsthand observations of rocks that have been rubbed together and washed to describe how they formed.
- **Engaging in argument from evidence** about the properties that make rocks belong to one group or another.
- **Obtaining, evaluating, and communicating information** about rocks and their properties.

NGSS Foundation Box for SEP

- **Ask questions** based on observations to find more information about the natural world.
- **Make observations** (firsthand) to collect data that can be used to make comparisons.
- **Record information** (observations, thoughts, and ideas).
- **Use and share pictures, drawings,** and/or writings of observations.
- **Use observations** (firsthand or from media) to describe patterns and/or relationships in the natural world in order to answer scientific questions.
- **Make observations** (firsthand or from media) to construct an evidence-based account for natural phenomena.
- **Construct an argument** with evidence to support a claim.
- **Read grade-appropriate text** and/or use media to obtain scientific and/or technical information to determine patterns in and/or evidence about the natural and designed world(s).
- **Obtain information** using various texts, text features, and other media that will be useful in answering a scientific question.

Exposing Crosscutting Concepts (CC)

NGSS Foundation Box for CC

- **Patterns:** Patterns in the natural and human designed world can be observed, used to describe phenomena, and used as evidence.
- **Cause and effect:** Events have causes that generate observable patterns; simple tests can be designed to gather evidence to support or refute student ideas about causes.
- **Stability and change:** Some things stay the same while other things change. Things may change slowly or rapidly.

Basalt
Bubble
Color
Data
Dull
Earth material
Flat
Geologist
Granite
Group
Mineral
Pattern
Pointed
Property
Rock
Rough
Round
Sand
Scoria
Shape
Sharp
Shiny
Size
Smooth
Sort
Texture
Tuff
Volcano
Weathering

The third dimension of instruction involves the crosscutting concepts, sometimes referred to as unifying principles, themes, or big ideas, that are fundamental to the understanding of science and engineering.

These concepts should become common and familiar touchstones across the disciplines and grade levels. Explicit reference to the concepts, as well as their emergence in multiple disciplinary contexts, can help students develop a cumulative, coherent, and usable understanding of science and engineering. (National Research Council, 2012, page 83)

In this investigation, the focus is on these crosscutting concepts.

- **Patterns.** Geologists look for and organize rocks by the patterns in the properties they observe. Patterns help to categorize earth materials and provide information about their formation.
- **Cause and effect.** Weathering (rubbing) causes large rocks to break into small rocks. Rocks of different sizes can be compared and described.
- **Stability and change.** Some rocks change quickly when rubbed and washed and some change very little. The rate of Earth-changing processes can be compared—some happen slowly and some rapidly.

Connections to the Nature of Science

- **Science addresses questions about the natural and material world.** Scientists study the natural and material world.

Conceptual Flow

The anchor phenomenon for this module is earth material—the solid rock and liquid water that we observe on Earth's surface. The guiding question for this module is what are properties of rocks and how do they change?

The **conceptual flow** for this first investigation starts with students observing **rocks** to describe their **properties** and the phenomenon of how they change. In Part 1, students are introduced to the first set of six rocks, composed of three types of volcanic rocks. Students focus on grouping rocks by type and then **rub** similar rocks together followed by rubbing different kinds of rocks together. They find that by rubbing rocks together, they weather rocks and get sand. The color and amount of the sand depend on the kind of rock. **Weathering** causes rocks to **break** into smaller pieces. Students study rocks like a **geologist**, using hand lenses, and generate a list of words to describe rocks. Students are introduced to the slowness of the weathering process and compare that to the rapid changes caused by volcanic activity.

In Part 2, students continue to study the three kinds of **volcanic rock** and learn the names for them—**basalt**, **tuff**, and **scoria**. They get the rocks wet and observe the **change in appearance** of the rock and what happens to the water. They observe air bubbles coming from some of the rocks, evidence of air pockets in the rock sample.

In Part 3, students investigate a new set of random rocks (**river rocks**) and practice using words to describe the properties of these new rocks. A **property** is any observable feature or characteristic of an object, in this case, a rock. Students learn that properties are the things we know about objects by observing them or feeling them. Students **group rocks by property** to develop their understanding of **rock properties**.

In Part 4, students go beyond the limits of the classroom to look for rocks in the schoolyard environment. After selecting two or three small rocks of interest, students share their discoveries by identifying one or more of their properties—**size or shape, color, texture, and pattern**. The **schoolyard rocks** start the class **rock collection**.

In Part 5, students use their understanding of rock properties (**size**, **shape**, **pattern**, **texture**, **color**) to engage in sorting activities with a partner. They also read about **granite** and study a sample to look for the colored **minerals** that make up the rock.

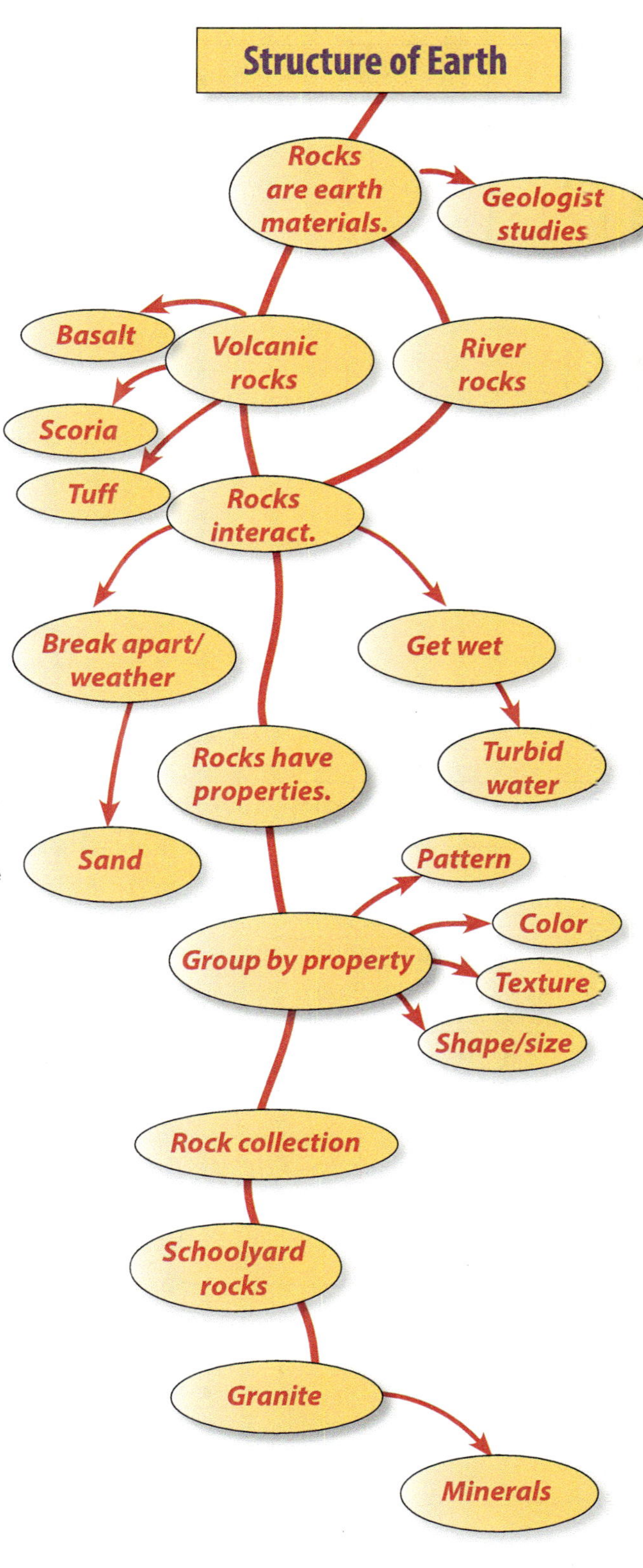

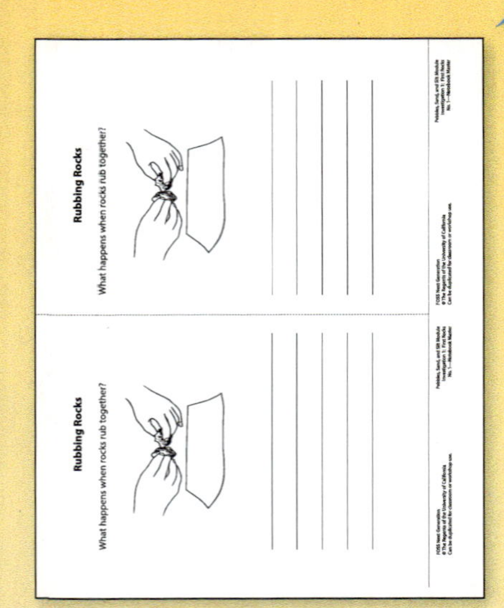

No. 1—Notebook Master

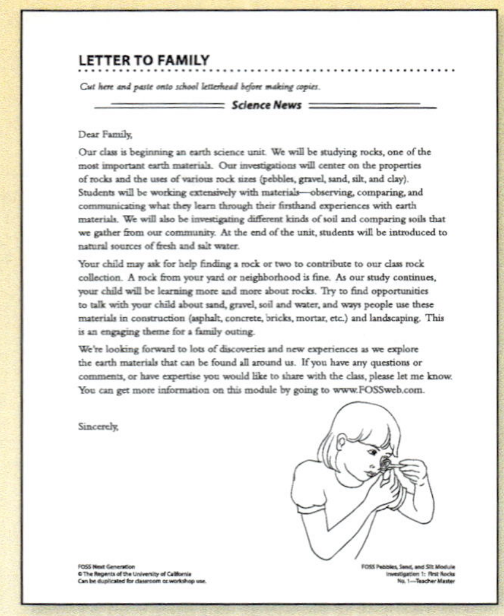

LETTER TO FAMILY

Cut here and paste onto school letterhead before making copies.

Science News

Dear Family,

Our class is beginning an earth science unit. We will be studying rocks, one of the most important earth materials. Our investigations will center on the properties of rocks and the uses of various rock sizes (pebbles, gravel, sand, silt, and clay). Students will be working extensively with materials—observing, comparing, and communicating what they learn through their firsthand experiences with earth materials. We will also be investigating different kinds of soil and comparing soils that we gather from our community. At the end of the unit, students will be introduced to natural sources of fresh and salt water.

Your child may ask for help finding a rock or two to contribute to our class rock collection. A rock from your yard or neighborhood is fine. As our study continues, your child will be learning more and more about rocks. Try to find opportunities to talk with your child about sand, gravel, soil and water, and ways people use these materials in construction (asphalt, concrete, bricks, mortar, etc.) and landscaping. This is an engaging theme for a family outing.

We're looking forward to lots of discoveries and new experiences as we explore the earth materials that can be found all around us. If you have any questions or comments, or have expertise you would like to share with the class, please let me know. You can get more information on this module by going to www.FOSSweb.com.

Sincerely,

No. 1—Teacher Master

MATERIALS *for*

Part 1: *Three Rocks*

For each student

- 1 Set of rocks
  - 2 Pieces of dark basalt
  - 2 Pieces of reddish scoria
  - 2 Pieces of light-colored tuff
- 1 Zip bag, 1 L
- 1 Half sheet of black paper (See Step 6 of Getting Ready.)
- 1 Half sheet of white paper ★
- 1 Hand lens
- 1 Science notebook (composition book) ★
- ❑ 1 Notebook sheet 1, *Rubbing Rocks*
- 1 *Letter to Family*

For each group

- 1 Paper plate

For the class

- 1 Set of rocks
- 2 Loupes/magnifying lenses
- 1 Paper plate
- 1 Half sheet of black paper
- 1 Vial with cap
- • Glue sticks ★
- • *Science Safety*, *Outdoor Safety*, and *Conservation* posters
- ❑ 1 Teacher master 1, *Letter to Family*

For embedded assessment

- ❑ • *Embedded Assessment Notes*

★ Supplied by the teacher. ❑ Use the duplication master to make copies.

GETTING READY *for*

Part 1: *Three Rocks*

1. Schedule the investigation

This part will take one active investigation session for observing rocks followed by a short notebook session. It can be done in one long session.

2. Preview Part 1

Students investigate and sort a set of six rocks. They gather information about the rocks by observing and comparing, then rub them together to simulate weathering. The focus question is **What happens when rocks rub together?**

3. Plan student organization

This activity is designed for groups of four to six students seated around tables or desks. Within the groups students work individually. Working in groups allows students to observe others, compare, share, and cooperate.

4. Set up a materials station

Plan to organize the materials for all the investigations in a location where one or two Getters from each group can get the materials efficiently. Assigning the role of Getter to students develops a sense of responsibility and decreases traffic flow in the classroom.

5. Prepare rock sets

Each student will have a set of six rocks. A rock set includes two pieces each of basalt, scoria, and tuff in a 1-liter (L) zip bag. Make one set for each student and one set for yourself. Decide whether to have Getters pick up the bags or to distribute them to each group in a basin.

6. Prepare half sheets of paper

Each student will need a half sheet of black paper and a half sheet of white paper. Cut the sheets in half. These sheets can be reused by the next class.

7. Set up an observation center

Set up a magnification center where students can view materials used in this and subsequent activities in the module. Place the two loupes/magnifying lenses and about four hand lenses at the center.

8. Plan to use a word wall or pocket chart

As the module progresses, you will add new vocabulary words to cards or sentence strips for use in a pocket chart and/or to a word-wall chart for posting on a wall or an easel.

TEACHING NOTE

The Getting Ready section for Part 1 of Investigation 1 is longer than the corresponding section for the other parts. Several of the numbered items appear only here, but may apply to other parts as well, such as the discussions about setting up a materials station (4), planning a word wall (8), preparing student science notebooks (10), and photocopying duplication masters (12).

NOTE

To prepare for this investigation, view the teacher preparation video on FOSSweb.

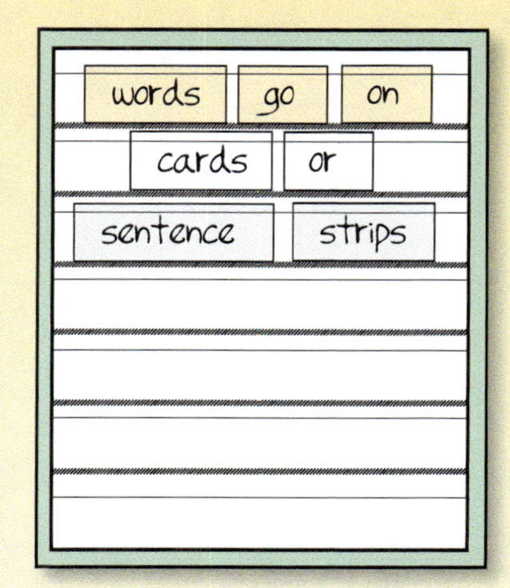

WORD WALL

words
words
and more words

EL NOTE

Display the equipment photo card for each object, and write the object's name on the word wall. Equipment photo cards are available on FOSSweb and can be downloaded and printed.

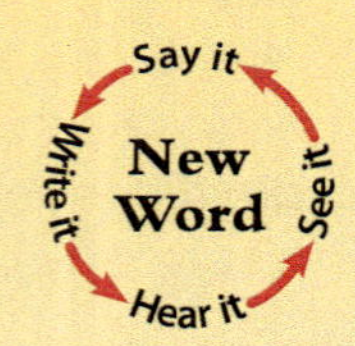

9. Plan for working with English learners

At important junctures in an investigation, you'll see a sidebar note titled "EL Note." These notes suggest additional strategies for enhancing access to the science concepts for English learners. Refer to the Science-Centered Language Development chapter for resources and examples to use when working with science vocabulary, writing, oral discourse, and readings in each investigation.

Each time new science vocabulary is introduced, you'll see the new-word icon in the sidebar. This icon lets you know not only that you'll be introducing important vocabulary, but also that you might want to plan on spending more time with those students who need extra help with the vocabulary.

10. Plan for student notebooks

Students will keep records of their science investigations in their science notebooks. They will record observations, responses to focus questions, and connections between their classroom learning and the world beyond. These records will be useful reference documents for students and revealing testaments of each student's learning progress for adults.

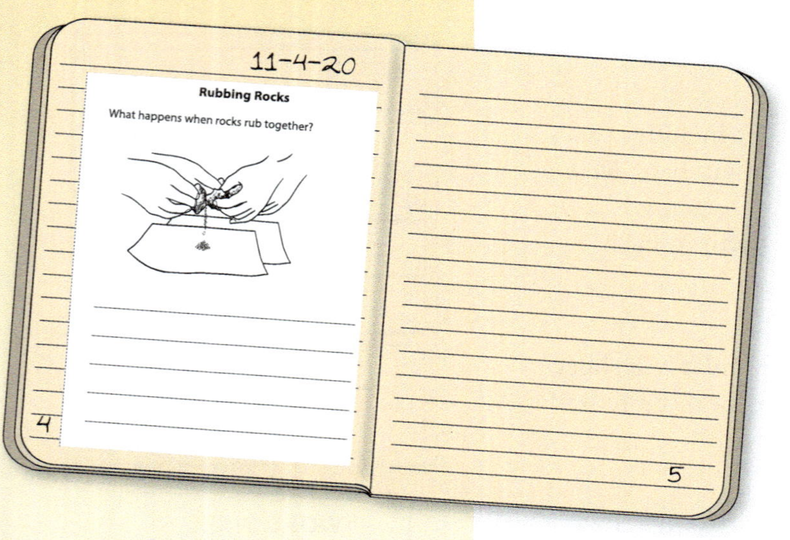

We recommend that students use bound composition books for their science notebooks. This ensures that student work becomes an organized, sequential record of student learning. Students' notebooks will be a combination of structured sheets photocopied from notebook masters provided in this teacher guide and unstructured student-generated drawings and writings on blank pages. In addition, student work can be entered partly in space provided on the notebook sheets and partly on adjacent blank sheets. Primary students with limited experience in organizing their own work may benefit from this approach.

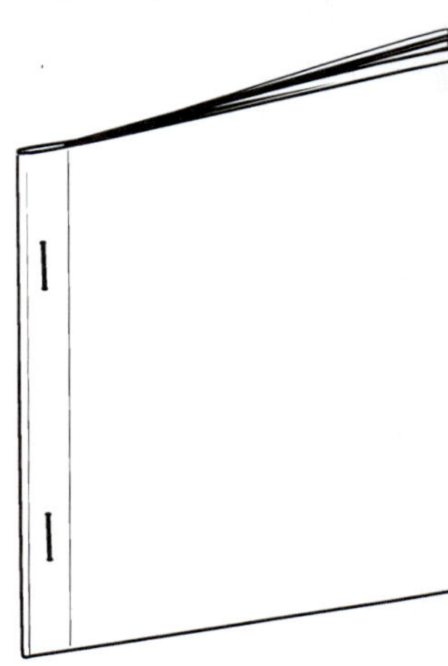

If you are already using an alternative method of organization with your students, such as a sheaf of folded and stapled pages, your method can take the place of the bound composition book.

11. Project images of notebook sheets

Throughout the module, you will see suggestions for projecting images of notebook sheets to orient students to their use. If you choose to use this support, use whatever technology is available to you. This may include a document camera, LCD projector and computer, or interactive whiteboard. Digital versions of all duplication masters for projecting are available on FOSSweb.

12. Print or photocopy duplication masters

Two sets of duplication masters are used throughout this module: notebook masters and teacher masters. Notebook masters are sized to be glued into a composition book. Each notebook master has two copies of the sheet, which should be cut apart so that each student gets a half sheet.

Teacher masters serve various functions—letter to family and take-home projects throughout the module.

A notebook sheet or teacher master that requires printing or duplication is flagged with this icon ❑ in the materials list for the part. It is referred to as a notebook sheet in the materials list and a notebook master in the sidebar snapshot.

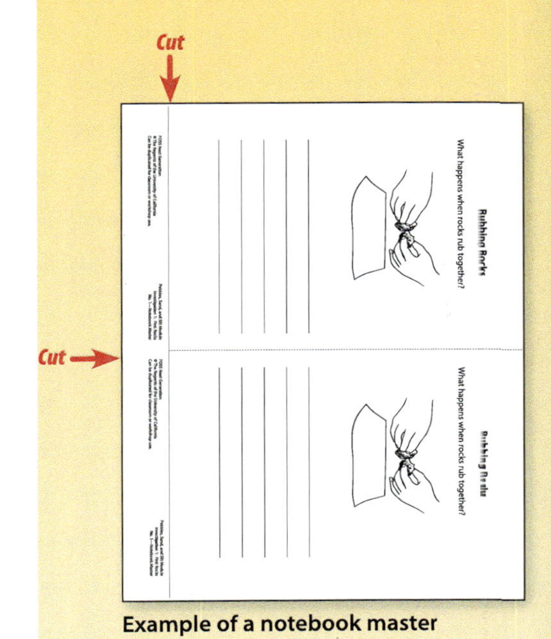

Example of a notebook master

13. Send a letter home to families

Teacher master 1, *Letter to Family*, is a letter you can use to inform families about this module. The letter states the goals of the module and suggests some home experiences that can contribute to students' learning.

14. Plan for safety and conservation

Primary students must be allowed to demonstrate that they can act responsibly with materials, but they must be given guidelines for safe and appropriate use of materials. Work with students to develop those guidelines so that students participate in making behavior rules and understand the rationale for the rules. Encourage responsible actions toward other students. Display and discuss the *Science Safety* and *Outdoor Safety* posters in class.

Look for the safety-note icon in the Getting Ready sections, which will alert you to safety concerns throughout the module. Be aware of any allergies your students have. When students rub rocks together, a small amount of rock dust comes off the rocks (sand). Be aware of students who have severe breathing problems and who might have trouble with rock dust. Caution students not to rub their eyes with sandy hands. Remind students to wash their hands with soap and water after handling materials.

TEACHING NOTE

This is a good time to introduce the set of four Conservation *posters and and discuss the importance of natural resources with students.*

15. Check FOSSweb

FOSSweb (www.FOSSweb.com) is the official FOSS website, with interactive instructional activities and media connections for use at school and at home. It has module updates and useful suggestions as well as many resources from the teacher guide that will help you plan the investigations. Become familiar with the resources on FOSSweb and plan to visit it often as you teach the module.

16. Plan for online activities

There will be opportunities to project online activities and streaming videos through a computer for the class to see. The Getting Ready section for each part will indicate what to prepare.

17. Assess progress throughout the module

Embedded (formative) assessments provide a variety of ways to gather information about students' thinking while their ideas are developing. These assessments are meant to be diagnostic. They provide you with information about students' learning so that you know whether to go on to the next part of the investigation or if you need to plan a next step to clarify understanding before going on to the next part. Most Getting Ready sections describe an embedded-assessment strategy that you may find useful in that part. Assessment master 1, *Embedded Assessment Notes,* is a half sheet provided to help you analyze student work. (See the Assessment chapter for more on how to use this sheet and the 10-minute reflective-assessment practice.) In some parts, the embedded assessment involves scientific practices. Assessment masters 2–3, *Performance Assessment Checklist,* is used for recording these observations.

Embedded Assessment Notes — Pebbles, Sand, and Silt Module

Investigation ___, Part ___ Date ___
Got it!
Concept
Concept
Reflections/Next Steps

Investigation ___, Part ___ Date ___
Got it!
Concept
Concept
Reflections/Next Steps

No. 1—Assessment Master

At the end of each investigation, students take an I-Check benchmark assessment. The items on these assessments examine all the concepts students have learned up to that point in the module. You can find out more about benchmark assessments in the Assessment chapter. Record I-Check data on assessment masters 4–5—*Assessment Record.*

18. Plan assessment: notebook entry

In Step 13 of Guiding the Investigation, students complete their first notebook entry (notebook sheet 1, *Rubbing Rocks*). Depending on when you are teaching this module in the school year, you may want to scaffold this entry for the class. Review students' work to make sure they know that smaller rocks come from bigger rocks (when they are rubbed together). (This is one of the processes that contributes to the slow but constant changes to Earth's surface.)

GUIDING *the Investigation*

Part 1: *Three Rocks*

1. Introduce the investigation

Call students to the rug. Hold up a bag of six rocks. Identify the objects as **rocks**. Ask students to tell a partner what they know about rocks. For example, how are rocks different? Where do they come from?

Tell them that rocks come from Earth so they are called **earth materials**. Explain,

Each of you will get a bag of rocks like this. Empty the rocks out on the table, observe them, and figure out how many different kinds of rocks there are. You can use a hand lens to observe the rocks up close.

2. Organize groups

Organize students into groups of four to six students. Assign one student from each group to be the Getter.

3. Distribute rocks

Distribute the rocks yourself or have Getters get a bag of rocks for each student in their group. Distribute one hand lens to each student. Allow 5–10 minutes for observing and sorting the rocks.

4. Discuss results (data)

Call students back to the rug, but have them leave the rocks at their tables. Ask students to report how many different kinds of rocks they found. Confirm that there are three different kinds of rocks, two of each type.

Hold up each kind of rock, one at a time, and ask students to offer words that describe the rock. Record their words on the word wall. Those words may include **size** (large, small), **texture** (**rough**, **smooth**), **color** (red, tan, gray), and **shape**. Explain that by recording these observations, we are collecting **data** on the rocks.

5. Focus question: What happens when rocks rub together?

After students have described the different kinds of rocks, ask them how they could find out what happens when rocks rub against each other. Listen to a few ideas, then project or write the focus question on the board, saying it aloud as you do.

➤ *What happens when rocks rub together?*

Suggest that students find out more about the different rocks by rubbing them together. To demonstrate, rub together one pair of

FOCUS QUESTION

What happens when rocks rub together?

TEACHING NOTE

The focus question in each part engages students with the phenomenon to investigate.

TEACHING NOTE

If this is the first module of the year, explain to students the role of the Getter.

Materials for Step 3

- *Bags of rocks*
- *Hand lenses*

TEACHING NOTE

Pulling students' attention away from materials can be difficult. Establish a signal with students that includes placing their hands on their chairs or their heads. Gathering students at the rug is another way to focus attention.

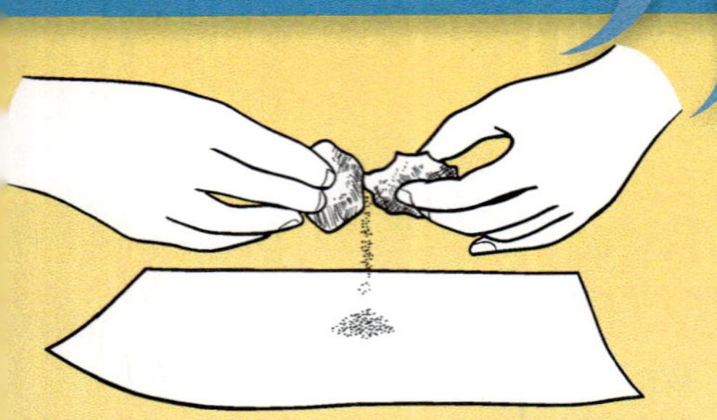

the same kind of rocks over a sheet of black paper. Suggest rubbing each pair of rocks over black and white paper and observing what, if anything, happens.

Tell students that it doesn't matter in what order they rub the rocks together, but suggest they first rub the same kind of rocks together.

Materials for Step 6
- *White paper*
- *Black paper*

6. Rub rocks together

Distribute a piece of black paper and a piece of white paper to each student. Have students begin rubbing the rocks.

Students will see tiny rock pieces break off and fall to the paper. They will probably notice that some of the rocks are harder than others, that some rocks leave marks on other rocks, and that they can write on the paper with some rocks.

Materials for Step 7
- *Paper plates*
- *Vial*

7. Collect the tiny rock pieces and rocks

Ask the Getters to get a paper plate for their groups. Have each student dump his or her tiny rock pieces on the plate. Collect all the rock debris on one paper plate. Transfer the pieces to a vial and show it to students. Explain that you are going to save the pieces for them to observe again later. Ask students to return the rocks to the zip bags. Have the Getters return the materials to the materials station.

Put some of the rock debris at the magnification center for students to view later.

SCIENCE AND ENGINEERING PRACTICES

Planning and carrying out investigations

CROSSCUTTING CONCEPTS

Cause and effect

8. Have a sense-making discussion

Call students back to the rug to discuss what they observed. A pocket chart is particularly useful for developing specific vocabulary as student offer descriptions.

Explain that people who study rocks are called **geologists**. Tell the class that they have worked like geologists as they found out about their rocks. Ask students to share their observations. You might use these questions to guide the discussion.

- ➤ *What happened when you rubbed two rocks together?* [Tiny pieces of rock broke off; one rock left a mark on the other rock.]
- ➤ *What should we call the tiny pieces of rock that came off the rocks?* [Sand or rock dust.]
- ➤ *Does each kind of rock change in the same way?* [Some change very little and remain the same; others change quickly.]
- ➤ *Was there one rock that made rock dust easier than the others?* [Tuff.]
- ➤ *Does the dust look the same on the white and the black paper?* [No.]
- ➤ *How does the dust feel between your fingers?* [Gritty.]

9. Introduce *weathering* and *sand*

Tell students,

*When rocks rub or bang against one another, small pieces can break off. This is called **weathering**. Weathering causes rocks to break into smaller pieces. The small pieces of rock that feel rough and gritty are **sand**.*

Add *geologist*, *weathering*, and *sand* to the word wall. You might also add illustrations to go with each word.

If necessary, clarify the meaning of the word *weathering* as it is used in science. Discuss how it is not the same as *whether*. Ask students why they think this word is used to describe the breaking down of rocks.

Ask students to focus on cause and effect.

- ➤ *What is the effect of rubbing the rocks?* [The rock gets a tiny bit smaller as bits of rock fall off. The tiny pieces are sand.]
- ➤ *What causes the tiny pieces of rock to come off?* [Weathering—the rubbing of the rocks together or the scraping of one rock against the other.]
- ➤ *Think about rocks you've seen outdoors. What does our investigation make you think about?*
- ➤ *How might rocks rub against each other in nature?* [Rolling down a hill in a rock fall.]
- ➤ *Do you think weathering happens quickly or slowly over time?* [We can rub rocks together in a short period of time, but we get just a little bit of sand; in nature, the process to get lots of sand would take a long time.]

10. Introduce science notebooks

Tell students that geologists and other scientists keep a notebook to write about their science discoveries. Every student will have his or her own notebook.

Distribute composition books to students. Give them a minute to confirm that the pages are all blank. Have students write the numbers 1–10 in the outside corner of the first ten pages.

NOTE: Writing the numbers on the pages is optional for primary students. This could be done during language-arts time to reinforce book format.

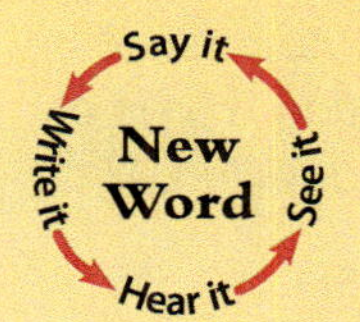

SCIENCE AND ENGINEERING PRACTICES

Analyzing and interpreting data

Constructing explanations

CROSSCUTTING CONCEPTS

Stability and change

NOTE

Go to FOSSweb for *Teacher Resources* and look for the Crosscutting Concepts—Grade 2 chapter for details on how to engage young students with this concept.

TEACHING NOTE

For this first notebook record, you might find it necessary to have everyone write the same response so you can model how to make a notebook entry by writing words or a sentence under the drawing.

EL NOTE

When steps are presented orally, consider writing or drawing them on the board and modeling the procedures. Look for procedural words that are important for following the directions. A list of these words is in the Science-Centered Language Development chapter.

TEACHING NOTE

For students writing directly in their notebooks, encourage them to use cause-and-effect statements such as When _____, we observed _____.

11. Attach the first notebook sheet

When the pages are numbered, call for attention. Restate the focus question and point to it on the board.

➤ *What happens when rocks rub together?*

If students are new to notebooks, tell them that you have a sheet with the question written on it. Distribute notebook sheet 1, *Rubbing Rocks*, to each student. Describe and model how to glue the sheet into the notebook.

a. *Open your notebook to page 4.*

b. *Write the date on the top of the page.*

c. *Use a glue stick to put glue on the back of the sheet.*

d. *Stick the sheet below the date.*

When students have their notebook sheets glued in place, read the focus question aloud together.

Or, if students are experienced using notebooks, simply have them date the page and record the focus question directly in their notebooks.

12. Answer the focus question

Ask,

➤ *What happens when rocks rub together?* [Small pieces break off, or sand falls off.]

Encourage students to use their own words on notebook sheet 1, *Rubbing Rocks*, and to add to the picture or draw a picture of their own. They can also use the blank page on the right to add more to their responses or tape small pieces of sand.

13. Assess progress: notebook entry

Collect students' notebooks after class and review their work.

What to Look For

- *Students recorded their observations by completing the picture to show small pieces of rock falling from the rocks to the table.*
- *The effect of rubbing rocks together is that "tiny pieces break off" or "it makes sand."*
- *When rocks rub together, it can change the shape of some rocks quickly; others can change more slowly.*

14. Review vocabulary

Review key vocabulary added to the word wall earlier. One way to do this is to use cloze review. You say a sentence, leaving the last word off, and ask students to answer chorally. Here's an example of a cloze review for this part.

➤ *A scientist who studies rocks is a _____.*

S: Geologist.

➤ *We are geologists because we are studying _____.*

S: Rocks.

➤ *When some rocks are rubbed together, they make _____.*

S: Sand.

➤ *Rocks break into smaller and smaller pieces because of _____.*

S: Weathering.

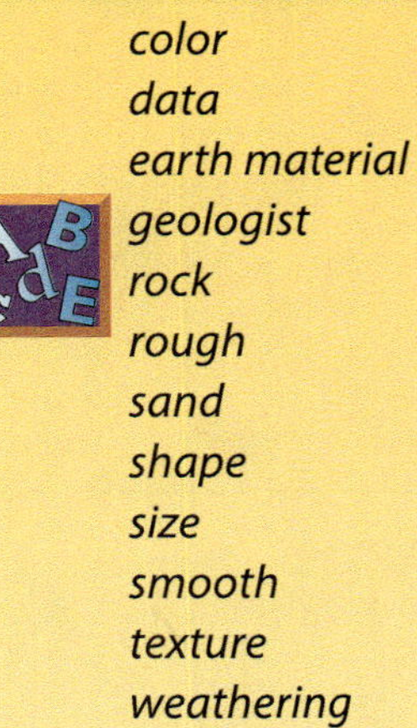

color
data
earth material
geologist
rock
rough
sand
shape
size
smooth
texture
weathering

EL NOTE

Write the cloze sentences on chart paper and give some students the vocabulary words on sentence strips. Have students hold up the appropriate word when the sentence is read.

WRAP-UP/WARM-UP

15. Share notebook entries

Conclude Part 1 or start Part 2 by having students share notebook entries. Ask students to open their science notebooks to the most recent entry. Read the focus question together.

➤ *What happens when rocks rub together?*

Ask students to pair up with a partner to

- share their answers to the focus question;
- share their drawings.

Encourage students to think about and discuss the cause-and-effect relationships involved in weathering. When we rubbed rocks together, the effect was we made sand or smaller pieces of rock. Another effect was the original rocks got a little smaller. Students can also discuss what changed and what stayed the same. The rock changed in size but is still made of the same materials.

ELA CONNECTION

This suggested strategy addresses the Common Core State Standards for ELA.

SL 2: Recount or describe key ideas.

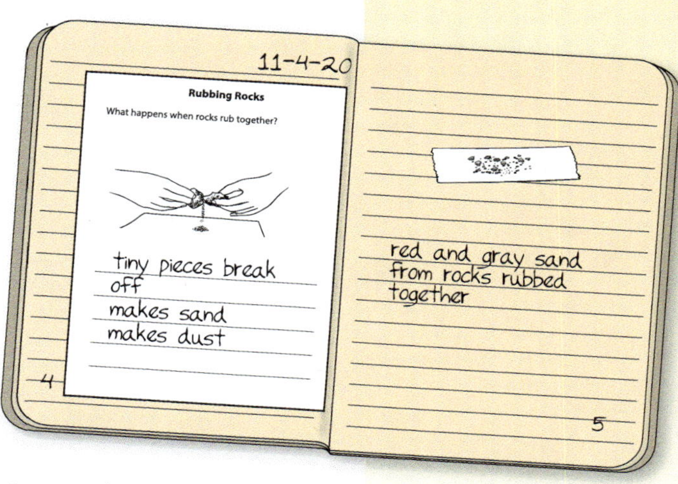

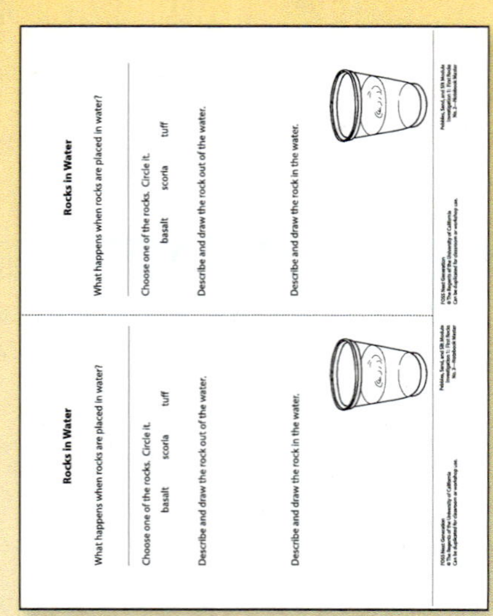

Rocks in Water

What happens when rocks are placed in water?

Choose one of the rocks. Circle it.

basalt scoria tuff

Describe and draw the rock out of the water.

Describe and draw the rock in the water.

Rocks in Water

What happens when rocks are placed in water?

Choose one of the rocks. Circle it.

basalt scoria tuff

Describe and draw the rock out of the water.

Describe and draw the rock in the water.

No. 2—Notebook Master

MATERIALS *for*

Part 2: *Washing Three Rocks*

For each student

- 1 Bag of six rocks (from Part 1)
- 1 Hand lens
- ❑ 1 Notebook sheet 2, *Rocks in Water*

For each pair of students

- 1 Plastic cup
- 1 Paper towel ★

For the class

- 1 Bag of six rocks
- 1 Basin
- 1 Pitcher
- • Paper towels ★
- • Water ★
- • Dry, powdered clay (optional)
- 1 Computer with Internet access ★

For embedded assessment

- ❑ • *Embedded Assessment Notes*

★ Supplied by the teacher. ❑ Use the duplication master to make copies.

GETTING READY *for*

Part 2: *Washing Three Rocks*

1. Schedule the investigation

This part will take one active investigation session for washing rocks followed by a short notebook session. It can also be done in one long session.

2. Preview Part 2

Students wash their rocks to see how they change when they get wet and to see what happens to the wash water. Students are introduced to the names of these volcanic rocks (tuff, scoria, basalt) and view a video on volcanoes to find out how they formed. The focus question is **What happens when rocks are placed in water?**

3. Plan for water

You will need to provide one plastic cup half-filled with water for each pair of students. You could provide a cup of water for each student but this is a good opportunity to have students practice sharing and working together. Fill a pitcher with water and have plenty of paper towels handy for wiping up spills.

4. Plan for storing rock sets

At the end of this part, dry the rocks thoroughly. This can be done by setting them out on a counter overnight to dry.

Rebag the dry rocks in sets of six. This is an excellent job for a few students. If the rocks are clean as a result of the activity, add a large pinch of dry powdered clay to each bag before storing them in the kit. This will provide some rock dust for the next time the rocks are washed. Powdered clay is included in the kit for this purpose.

5. Preview the videos

Preview the video clip of stream water flowing over rocks (duration 42 seconds, from the interactive eBook, page 7). Show this at the end of Step 8. Also preview the video, *All about Volcanoes* (duration 21 minutes). The video provides visuals of Earth events that happen quickly to contrast with weathering events that take a long time. Determine a good pausing location after the first 3 minutes of video (see Step 14 of Guiding the Investigation). The links to these videos are in the Resources by Investigation on FOSSweb.

TEACHING NOTE

We recommend that students watch the whole video to engage with the visuals.

6. Plan assessment: notebook entry

In Step 11 of Guiding the Investigation, students answer the focus question. Students answer a second question in Step 12. Collect students' notebooks after class and check to see if students connect washing rocks with movement of the small particles.

FOCUS QUESTION

What happens when rocks are placed in water?

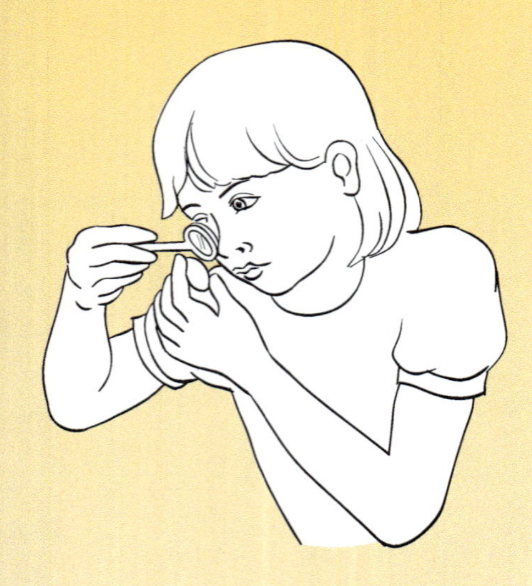

Materials for Steps 4–5

- *Bags of rocks*
- *Cups*
- *Pitcher of water*
- *Hand lenses*
- *Paper towels*

SCIENCE AND ENGINEERING PRACTICES

Planning and carrying out investigations

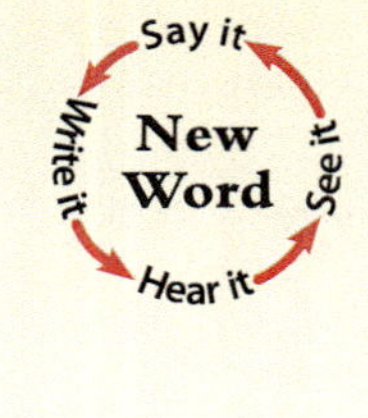

GUIDING *the Investigation*

Part 2: *Washing Three Rocks*

1. Describe proper use of a hand lens

Call students to the rug. Now is a good time to encourage the proper use of the hand lens to observe fine detail on an object such as a rock. Show students how to hold the lens against their eye socket and then move the rock until it comes into focus.

2. Focus question: What happens when rocks are placed in water?

Ask students to think about how water affects rocks. Give them time to brainstorm questions they have about rocks and water. Suggest that they use the three kinds of rocks to find out what happens when they place the rocks in water. They can use the hand lens to observe the rocks before they go into the water, and again when they take them out. Project or write the focus question on the board.

➤ *What happens when rocks are placed in water?*

3. Explain the washing procedure

Tell the class that each student will get a bag of rocks and a hand lens; each pair of students will get a cup of water and a piece of paper towel. Describe the procedure.

a. *Choose one kind of rock from the bag and observe it dry.*

b. *Put only one rock at a time in the water.*

c. *Observe the wet rock carefully while it's in the cup.*

d. *Remove the rock from the water and place it on a paper towel.*

e. *Repeat the process with the other rocks.*

4. Distribute materials

Have students return to their tables. Have the Getters pick up the materials, including empty cups.

5. Start washing rocks

Provide each pair of students with a half-filled cup of water and a paper towel. Allow 10–15 minutes for rock washing. Move from group to group, asking these questions.

➤ *Did the rocks change when you put them in the water? How?* [Some change colors, some are **shiny**, some made **bubbles**.]

Students may get very excited when they see air bubbles coming off the scoria. Have them think about where the bubbles are coming from.

➤ *Is there anything you can see now that you couldn't see when the rocks were dry?* [**Dull** rocks showed colors, stripes or new **patterns**.]

➤ *What happened to the water after you put the rocks in?* [It got muddy.]

6. Dry the rocks

Have students place the rocks on a fresh paper towel to dry on their desks. They should wipe up any spilled water. The cups of water can remain until after the discussion.

7. Discuss results

Review the questions in Step 5. Have one set of the rocks and three cups of water for this review. Start by asking students to talk with a partner to describe what happened to each kind of rock when it was placed in water.

Then review what happened to the water. Give students time to talk with a partner.

8. Have a sense-making discussion

Demonstrate how to pour the water from the cups into a basin. Tell students this is what they will do. Distribute the basin to each group and have students pour the water off and observe the residue remaining in the cups. Ask them where that material came from. Reinforce that it came off the rocks.

Ask if students have any new words to describe rocks or observations. Add the new words to the class word wall.

After the water is in the basin, ask the students to think about how water might change earth materials or land over time. Ask,

➤ *What do you think happens to those tiny pieces of rock if they wash off a rock that falls into moving water, like a stream or a river?*

➤ *How might water cause the land to change over time?*

➤ *How long would it take for water to change the way land looks?*

Show the video clip of water rushing over rocks, an example of weathering—a slow process. Discuss student observations.

9. Introduce the rock names

Gather students together near the word wall. Hold up each kind of rock, one at a time. Have students describe it, and reveal its name. Add the rock names to the word wall.

Basalt *is the gray, hard rock. It was once hot liquid lava from a volcano.*

Tuff *is the light, soft rock. It is made of ash from the fires of the volcano.*

Scoria *is the reddish bubbly rock. It was once the bubbly top of the lava.*

TEACHING NOTE

The term patterns *here means the designs or markings on the rock. This is different than the crosscutting concepts of patterns.*

Materials for Step 6

- *Basin*
- *Paper towels*

SCIENCE AND ENGINEERING PRACTICES

Analyzing and interpreting data

CROSSCUTTING CONCEPTS

Stability and change

TEACHING NOTE

The flowing stream video clip is from the first article in the FOSS Science Resources *eBook. Students will discuss this again in Part 4 when they read the article. In Part 4, students develop the concept that weathered river rocks are smoother than other rocks.*

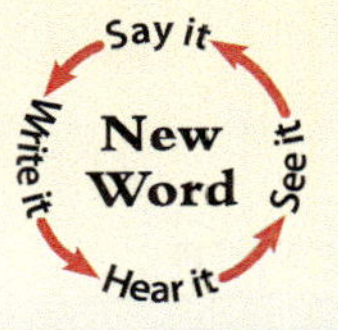

EL NOTE

Print the equipment photo card for each rock and place it next to the appropriate rock name on the word wall.

basalt
bubble
dull
pattern
scoria
shiny
tuff
volcano

SCIENCE AND ENGINEERING PRACTICES

Analyzing and interpreting data

SCIENCE AND ENGINEERING PRACTICES

Constructing explanations

Obtaining, evaluating, and communicating information

10. Review vocabulary

Review key vocabulary added to the word wall in this part. It will help students when they answer the focus question.

11. Answer the focus question

Ask the class,

➤ *What happens when rocks are placed in water?*

Distribute a copy of notebook sheet 2, *Rocks in Water*, to each student. Have them glue the sheets into their notebooks.

Tell them that they should pick one of the three rocks and answer the focus question for that rock. Have students who finish the task quickly complete a notebook entry for a second rock on a blank sheet in their notebooks.

12. Assess progress: notebook entry

Have students focus on the small pieces of rock that washed off in the water.

What to Look For

- *Students recorded their observations by drawing a picture of one of the dry rocks and writing a word to describe one thing about it.*
- *Students recorded their observations of effects caused by placing the rocks in water (change of color, observed air bubbles, etc.).*
- *Students explain that the tiny pieces of rock that wash off will most likely be carried downstream.*

13. Clean up

Recycle the washing water now in the basin. Use it to water plants indoors or outdoors.

If the rocks are dry, have students return them to the bags and bring them to the materials center. Otherwise, collect them at a drying station and have a group bag them later. If the rocks are clean as a result of the activity, add a large pinch of powdered clay to each bag before storing them in the kit.

14. View the video: *All about Volcanoes*

Ask students to share with a partner what they already know about **volcanoes**. Add volcano to the word wall. Show the first 2–3 minutes of the video without the sound and just let students observe and talk about what they observe in this "virtual field trip." Then ask them to generate questions as you make a list of these questions for the class.

Explain that volcanic eruptions are an example of an event that can change the land quickly. Tell students to look for evidence in the video that supports this claim. Have students watch the video from the start with the sound (lava spewing and flowing).

Here is a summary of the 21-minute video.

- Introduction
- Parts of a volcano
- Investigation: Building a model volcano
- Types of volcanoes (Students will revisit volcano types, cinder cones, shield, composite volcanoes, in the "Landforms" reading at the end of Investigation 2.)
- Active and dormant volcanoes
- Why we need volcanoes
- Predicting eruptions
- Conclusion

Discuss examples from the video of the land changing quickly.

15. Review the three volcanic rocks

Bring out the three types of volcanic rocks and review how they were formed focusing on cause and effect—pressure results in eruptions.

***Basalt** is the gray, hard rock. It was once hot liquid lava from a volcano.*

***Tuff** is the light, soft rock. It is made of ash from the fires of the volcano.*

***Scoria** is the reddish bubbly rock. It was once the bubbly top of the lava.*

CROSSCUTTING CONCEPTS

Cause and effect

Stability and change

WRAP-UP/WARM-UP

16. Share notebook entries

Conclude Part 2 or start Part 3 by having students share notebook entries. Ask students to open their science notebooks to the most recent entry. Read the focus question together.

➤ *What happens when rocks are placed in water?*

Ask students to pair up with a partner to

- share their answers to the focus question;
- share their drawings.

Encourage students to think about and discuss the changes they observed in terms of cause and effect. When we put the rocks in water, the effect was they changed color. Air that was in the holes in the rock caused air bubbles to come out underwater.

ELA CONNECTION

This suggested strategy addresses the Common Core State Standards for ELA.

SL 2: Recount or describe key ideas.

MATERIALS *for*

Part 3: *First Sorting*

For each pair of students

- 1 Set of 20 large pebbles (See Step 3 of Getting Ready.)
- 1 Zip bag, 1 L
- 2 Hand lenses
- 1 Plastic cup
- ❑ 2 Notebook sheet 3, *Rock Sorting*

For the class

- 1 Set of 20 large pebbles
- 1 Zip bag, 1 L
- 1 Basin
- 1 Pitcher
- • Paper towels ★
- • Water ★
- • Dry powdered clay

For embedded assessment

- ❑ • *Performance Assessment Checklist*

★ Supplied by the teacher.

❑ Use the duplication master to make copies.

Rock Sorting

How are river rocks the same?

I sorted the rocks into groups using the property of

Rock Sorting

How are river rocks the same?

I sorted the rocks into groups using the property of

No. 3—Notebook Master

GETTING READY *for*

Part 3: *First Sorting*

1. **Schedule the investigation**
   This part will take one active investigation session for sorting rocks followed by a short notebook session. It can also be done in one long session.

2. **Preview Part 3**
   Students are introduced to river rocks, describe their properties, and compare and sort them into groups based on one property at a time. The focus question is **How are river rocks the same?**

3. **Prepare river-rock sets**
   Put 20 large pebbles (from the kit) in a zip bag for each pair of students. Make a set for yourself as well.

   These river rocks will be "dirty" when you receive them. Students will wash them in this activity. At the end you will need to add a pinch of powdered clay to get the rocks "dirty" for the next class.

   If you have a rocky stream or beach close by, you may want to add variety by collecting 300–400 large pebbles 2–4 centimeters (cm) in diameter to use in this investigation. River rocks are also often sold at nursery and garden supply stores.

4. **Plan assessment: performance assessment**
   In Step 8 of Guiding the Investigation, students sort river rocks by a single property. Check to see how well students observe the rocks, explain how they sorted them, can argue for how they sorted the rocks. Carry the *Performance Assessment Checklist* with you as you visit the groups while they work. The *Performance Assessment Checklist* starts with assessment master 2.

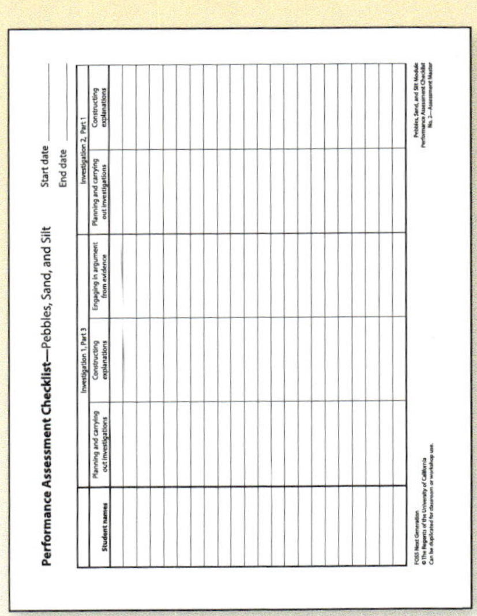

Performance Assessment Checklist—Pebbles, Sand, and Silt

Start date ______

End date ______

	Investigation 1, Part 3			Investigation 2, Part 1	
Student names	Planning and carrying out investigations	Constructing explanations	Engaging in argument from evidence	Planning and carrying out investigations	Constructing explanations

FOSS Next Generation

Pebbles, Sand, and Silt Module
Performance Assessment Checklist
No. 2—Assessment Master

No. 2—Assessment Master

FOCUS QUESTION

How are river rocks the same?

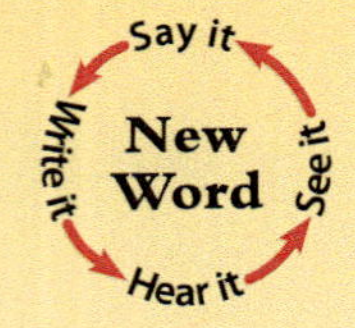

Materials for Step 4

- *Bag of large pebbles*

GUIDING *the Investigation*

Part 3: *First Sorting*

1. Review rocks

Call students to the rug and review what they have discovered about rocks so far.

2. Introduce *properties*

Tell students,

We have been using rock words to describe the rocks. Those words describe rock ***properties****. Things we know about rocks and other objects by looking at them or feeling them are their properties. Properties of rocks that you have discovered are rough, smooth, dull, shiny <and other properties observed by students>.*

Write *property* on the word wall.

3. Reinforce *properties*

One way to reinforce a new vocabulary word is to have students respond chorally. Below is an example of this mini-lesson.

➤ *I would like everyone to say "properties."*

S: Properties.

➤ *Let's clap the parts (syllables) of the word.*

S: Prop (clap) er (clap) ties (clap).

➤ *Again.*

S: Prop (clap) er (clap) ties (clap).

➤ *How many parts (syllables) are in the word properties?*

S: Three.

➤ *Properties are the things we know about objects by looking at them or feeling them. What are properties?*

S: Things we know about objects by looking at them or feeling them.

➤ *If something is rough, that is one of its _____.*

S: Properties.

➤ *Objects can be described by their _____.*

S: Properties.

4. Introduce river rocks

Bring out your bag of river rocks and say,

Let's continue our work as geologists. To find out more about these earth materials, we need to observe closely and look for patterns.

Show students some of the rocks and tell them,

Today I brought some different rocks to class. They are the kind of rocks you might find on the bottom of a river. We're going to call them river rocks.

Give students time to discuss with a partner how they think the river rocks might be the same or different from the volcanic rocks they studied earlier (basalt, tuff, scoria.)

5. Focus question: How are river rocks the same?

Introduce the focus question and project or write it on the board as you say it aloud.

➤ *How are river rocks the same?*

6. Introduce rock sharing

Explain that each pair of students will get a bag of river rocks. Tell them that they should divide the rocks so that each person gets a fair share. They should observe the rocks as they share them.

7. Distribute materials

Have students move to their tables. Ask the Getters to get one bag of rocks and two hand lenses for each pair of students in their group. Ask students to begin sharing the rocks and making observations.

8. Introduce sorting by a property

After each person has their river rocks, explain that their task is to **sort** the river rocks by a single property. Ask students to suggest properties they might use to sort the river rocks into **groups**.

As they offer ideas, write them on the board. Some of the properties they might offer include size (large/small), shape (**round**/**flat**), color, texture (rough/smooth), and pattern.

Challenge each pair of students to use just one property to sort the rocks. The two students in a pair will need to work together to sort the rocks based on that one property. Then they can select a second property and re-sort the rocks into new groups.

9. Monitor sorting

Monitor the pairs as they divide and sort the rocks into groups. Ask them to describe how they are sorting their rocks. Encourage students to sort the rocks using only one property at a time. If a pair has a problem getting started, suggest that they think about whether the rocks could be sorted by color. Allow about 10 minutes for the sorting.

Materials for Step 7

- *Bags of large pebbles*
- *Hand lenses*

TEACHING NOTE

After students divide the rocks, they often find it is more interesting to put all the rocks back together for sorting.

SCIENCE AND ENGINEERING PRACTICES

Planning and carrying out investigations

SCIENCE AND ENGINEERING PRACTICES

Planning and carrying out investigations

Constructing explanations

Engaging in argument from evidence

DISCIPLINARY CORE IDEAS

PS1.A: Structure and properties of matter

CROSSCUTTING CONCEPTS

Patterns

flat
group
pointed
property
round
sharp
sort

10. Assess progress: performance assessment

Monitor the pairs as they divide and sort the rocks into groups. Ask them to describe how they are sorting their rocks. Encourage students to sort the rocks using only one property at a time.

NOTE: If a pair has a problem getting started, suggest that they think about whether the rocks could be sorted by color.

What to Look For

- *Students make accurate observations of rocks and can sort them by property. (Planning and carrying out investigations; PS1.A: Structure and properties of matter; patterns.)*
- *Students can explain why they grouped the rocks the way they did. (Constructing explanations.)*
- *Students can argue for their grouping if challenged by the teacher. (Engaging in argument from evidence.)*

11. Discuss results and review properties

Call students to the rug to discuss what they observed and the different ways they sorted the rocks into groups. Review the property words on the word wall.

NOTE: You might look ahead to the Language section of the Interdisciplinary Extensions. At the bottom of the page you will see a graphic of a property map that might be useful to generate at this point in the activity.

12. Answer the focus question

Ask the class,

➤ *How are river rocks the same?*

Distribute a copy of notebook sheet 3, *Rock Sorting*, to each student. Have them glue the sheet into their notebooks.

Tell students that they should think about one of the ways that they sorted the rocks to answer the focus question. Have students who finish the task quickly complete a notebook entry for a second sorting property on a blank sheet in their notebook.

13. Suggest using water

Get students' attention. Ask them how water might be of use at this time. They will probably remember that the rocks changed color when placed in water. Tell them that each pair will use a plastic cup half full of water for washing the rocks. They should follow the same procedure as before, observing one rock at a time in the water.

14. Distribute water

Have the Getters pick up cups and paper towels for each student. Fill cups half full with water. As students place the rocks in water, ask about the changes they observe in both the rocks and the water.

Materials for Step 14

- *Cups*
- *Paper towels*
- *Pitcher of water*

15. Suggest re-sorting the wet rocks

Ask students to work with their partners to sort the rocks again now that they are wet.

16. Clean up

Have students pour the remaining water into a basin at the materials station and wipe up any spilled water. Collect the plastic cups.

Find a place where students can place the rocks to dry (it might take overnight). Make sure the rocks are thoroughly dry before returning them to the zip bags.

Materials for Step 16

- *Basin*
- *Paper towels*

WRAP-UP/WARM-UP

17. Share notebook entries

Conclude Part 3 or start Part 4 by having students share notebook entries. Ask students to open their science notebooks to the most recent entry. Read the focus question together.

➤ *How are river rocks the same?*

Ask students to work with a partner to

- share their answers to the focus question.

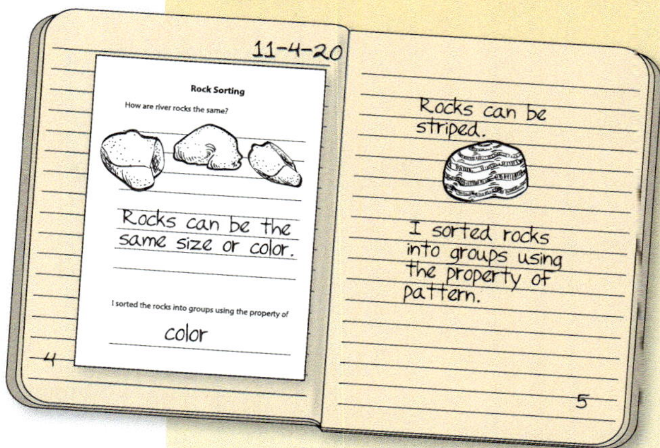

Encourage students to think about and discuss how looking for patterns helped them sort the rocks into groups.

You might use this as an opportunity to have students engage in argument from evidence. For example, say,

I heard someone say that river rocks aren't the same at all. They all look different. Do you agree or disagree?

Encourage students to provide evidence from their rock investigations to support how river rocks are the same or how they are different.

TEACHING NOTE

Go to FOSSweb for Teacher Resources *and look for the* Science and Engineering Practices—Grade 2 *chapter for details on how to engage second graders with the practice of engaging in argument from evidence.*

SCIENCE AND ENGINEERING PRACTICES

Engaging in argument from evidence

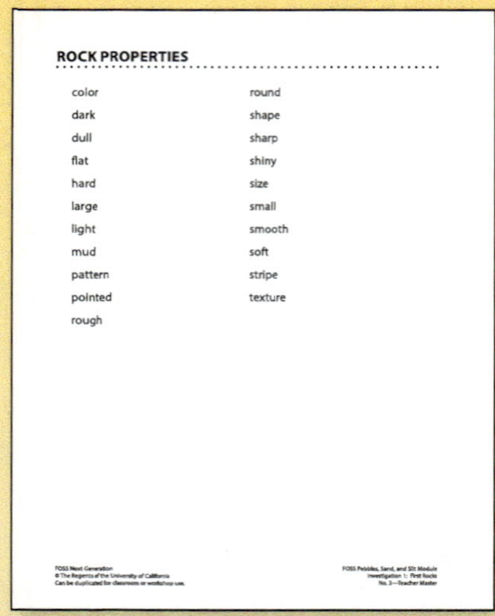

ROCK PROPERTIES

color	round
dark	shape
dull	sharp
flat	shiny
hard	size
large	small
light	smooth
mud	soft
pattern	stripe
pointed	texture
rough	

No. 2—Teacher Master

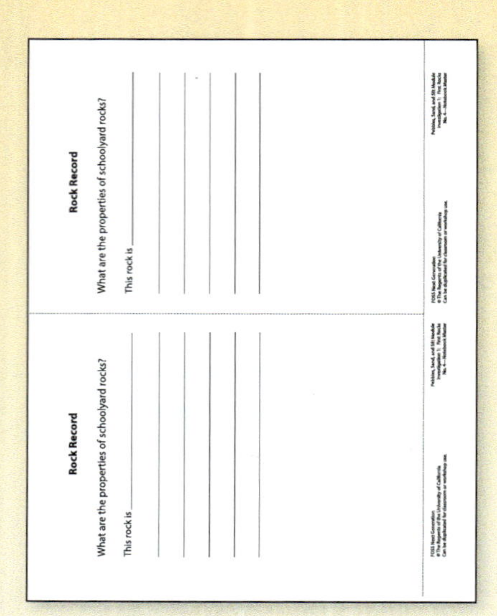

Rock Record

What are the properties of schoolyard rocks?

This rock is ________

Rock Record

What are the properties of schoolyard rocks?

This rock is ________

No. 4—Notebook Master

MATERIALS *for*

Part 4: *Start a Rock Collection*

For each student

- 1 Hand lens
- 1 Plastic cup
- ❑ 1–2 Notebook sheet 4, *Rock Record*
- 1 *FOSS Science Resources: Pebbles, Sand, and Silt*
  - "Exploring Rocks"

For the class

- 16 Plastic cups
- 1 Whistle or bell ★
- 1 *Outdoor Safety* poster
- • Water ★
- • Water containers (See Step 5 of Getting Ready.) ★
- 1 Pitcher
- 1 Bag for carrying materials ★
- • Egg cartons (optional) ★
- ❑ 1 Teacher master 2, *Rock Properties*
- 1 Big book, *FOSS Science Resources: Pebbles, Sand, and Silt*

For embedded assessment

- ❑ • *Embedded Assessment Notes*

★ Supplied by the teacher. ❑ Use the duplication master to make copies.

GETTING READY *for*

Part 4: *Start a Rock Collection*

1. Schedule the investigation

This part will take two active investigation sessions—one outdoor session to find rocks and one session for notebook rock records. Plan an additional session for the class reading.

2. Preview Part 4

Students take a field trip to collect and observe schoolyard rocks. They describe the properties of the various rocks. The focus question is **What are the properties of schoolyard rocks?**

3. Select your outdoor site

Survey your schoolyard for a site with rocks. You may need to walk around the school to several locations for students to find a variety to observe. Check pavement areas and walkways. Plan to end the walk in a location where students can sit down and wash their rocks. Refer to the Taking FOSS Outdoors chapter for more details on safety and strategies for working with students outdoors.

TEACHING NOTE

If you have trouble finding rocks in your schoolyard, you can collect them elsewhere and "seed" a location in the schoolyard with these rocks for students to find.

4. Prepare outdoor tools

Students will carry plastic cups outdoors to contain their rock collections. They can also carry hand lenses. Collect the hand lenses at the end of the outdoor session to minimize scratching of the lenses.

5. Plan for water

You will need 3–4 L of water for this outdoor activity. Plan how you can most easily convey water to your outdoor site. Two-liter soft-drink bottles or gallon milk jugs with caps are easy for small helpers to handle. A bucket is also suitable. The pitcher will also work but is harder to carry when it is full of water.

6. Check the site

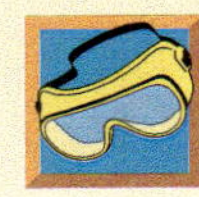

It is always a good idea to check the outdoor site on the morning of the outdoor activity. Check for anything that might be harmful or disrupt the outdoor lesson.

7. Set up a rock study center

Plan where you will display the rocks that students bring in. It should be where there is a large flat surface. Provide several hand lenses. Egg cartons can serve as containers for organizing and displaying the rocks. This center will become the home of a growing rock collection. Place a copy of teacher master 2, *Rock Properties,* at the center. This list can grow as students add words to the class word wall.

EL NOTE

You might make more copies of teacher master 2, Rock Properties, for individual students (see Step 11 of Guiding the Investigation).

8. **Plan to read *Science Resources*: "Exploring Rocks"**
Plan to read "Exploring Rocks" during a reading period after completing this part. Plan to show and discuss the video clip in the interactive eBook from page 7. This was the same clip students reviewed in Part 2, Step 8.

9. **Plan assessment: notebook entry**
Students use the *Rock Record* notebook sheet to record the properties of one schoolyard rock. Review these records to see if students can describe the properties of rocks in words and pictures.

GUIDING *the Investigation*

Part 4: *Start a Rock Collection*

1. Introduce rock collecting

Call students to the rug. Say,

We have been doing what geologists do—investigating rocks to determine their properties and observe how they change. Geologists often collect rocks from the local area to study and make rock collections.

Ask,

- ➤ *If you wanted to find some rocks to collect, where would you look?*
- ➤ *Do you think we have rocks in our schoolyard?*
- ➤ *Where might we find rocks in our schoolyard?*
- ➤ *What kinds of rocks do you think we might find in the schoolyard?*

2. Focus question: What are the properties of schoolyard rocks?

Remind students that they know the properties of some rocks from volcanoes and river rocks. Ask the focus question as you project or write it on the board.

- ➤ *What are the properties of schoolyard rocks?*

3. Describe the activity

Tell students that they will go outdoors to the schoolyard to search for as many as five small rocks (no more). Encourage students to search for rocks that have interesting properties. Ask,

- ➤ *What tools will help us collect and observe our rocks?*

Explain that each student can use a plastic cup to hold the rocks they find and a hand lens to observe the rocks. There will be water for rock washing after they collect rocks.

4. Discuss rock-collecting rules

Ask students to volunteer any rules they think they should follow when choosing rocks for the class collection. For example, the rocks shouldn't be too big or too valuable, or from someone else's yard unless they ask permission.

5. Review the rules for going outdoors

Review the procedures and behaviors for outdoor learning. Refer to the *Outdoor Safety* poster provided in the kit. In addition, highlight these guidelines.

- *Remember who your outdoor discovery buddy is.*
- *Walk quickly and quietly to the outdoor starting place.*

FOCUS QUESTION

What are the properties of schoolyard rocks?

TEACHING NOTE

*This is a good time to read the book **Everybody Needs a Rock** by Byrd Baylor. This trade book is available in most libraries.*

- *Start by forming a sharing circle—hands on hips, elbows touching. Wait for further instructions.*
- *When you hear the signal (whistle, bell, etc.), quickly return to the sharing circle.*

Materials for Step 6

- *Hand lenses*
- *Cups*
- *Containers of water*
- *Carrying bag*

6. Go outdoors

Have students line up in buddy pairs. Pick up the tools and water and walk outdoors in the usual orderly manner, using your outdoor learning door (not the door that is used to go to recess). Form a sharing circle and describe where the search will be conducted and how much time students will have to search for the rocks.

Remind them to collect no more than five rocks. Let the search begin. Allow students about 5–8 minutes.

7. Search for rocks

Circulate among pairs as they search for and collect their rocks. Ask questions such as,

➤ *What do you find interesting about that rock?*

➤ *Have you ever seen rocks like this one?*

➤ *Are there more rocks like this in our schoolyard?*

➤ *Do you have any rocks that are the same or that share a property?*

➤ *What questions do you have about the rocks that you found outside?*

SCIENCE AND ENGINEERING PRACTICES

Asking questions

8. Wash rocks outdoors

Alert students when they have 1 minute left. Give your signal for students to return to the sharing circle with their newly found rocks. Give each pair of students a cup of water and ask that they sit near each other to wash their rocks. Students may need a reminder (and sometimes a little help) to find a level spot to place their cups.

After washing their rocks, students may decide to discard some of their rocks and try others. Students should end the session with two or three rocks that they would like to add to a classroom rock collection.

9. Have a sense-making discussion

If some students are still collecting rocks, give your signal for students to return to the sharing circle with their rocks. Ask,

➤ *What are the properties of schoolyard rocks?*

➤ *Do schoolyard rocks have the same properties as river rocks?*

➤ *How does the texture of schoolyard rocks compare to river rocks?*

Ask students to observe their rocks and describe to their partners some of the rock properties. After a few minutes, call on a student to share one property of a rock he or she collected. Ask all those who found a rock with a similar property to hold it up. Focus on color, size, shape, and texture. Repeat the process several times.

10. Return to class

Have students select no more than three rocks to take back to the classroom. Return the other rocks to the schoolyard. Recycle any water still in cups. Line up, collect the hand lenses, and head back to the classroom in the usual orderly manner.

POSSIBLE BREAKPOINT

11. Answer the focus question

Back in the classroom, have students observe their schoolyard rocks with hand lenses. Ask them to complete a *Rock Record* sheet for at least one rock and glue the sheet into their notebooks.

For students who need help with property words, make copies of teacher master 2, *Rock Properties*.

Materials for Step 11

- *Hand lenses*
- *Schoolyard rocks*
- ***Rock Record** sheets*
- ***Rock Properties** sheets (optional)*

12. Assess progress: notebook entry

Review student work on notebook sheet 4, *Rock Record*.

What to Look For

- *Students recorded their observations by drawing a picture resembling a rock.*
- *Students described the properties of the rock they drew.*
- *Students draw their rocks to scale (drawing indicates size or rock).*

SCIENCE AND ENGINEERING PRACTICES

Analyzing and interpreting data

13. Display the rocks

Discuss how the class rock collection should be displayed. Suggest that students come up with ways to arrange the rocks in the display, such as by color, size, shape, or texture. After the class agrees on an arrangement, have a few students begin organizing the display. Call groups of students up to add their rocks to each category.

The rock collection can be reorganized in different ways while it exists in the classroom. Students may also bring in more rocks to add to the collection.

14. Expand the rock collection

The rock hunt will inevitably lead some students to discuss a fancy rock they found or have at home, such as a cut agate or a geode. You may want to suggest that they bring the rock to class one day to share, and then take the rock home for safekeeping.

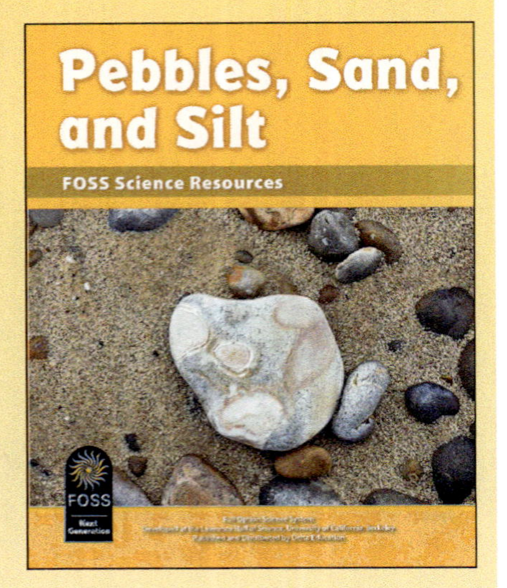

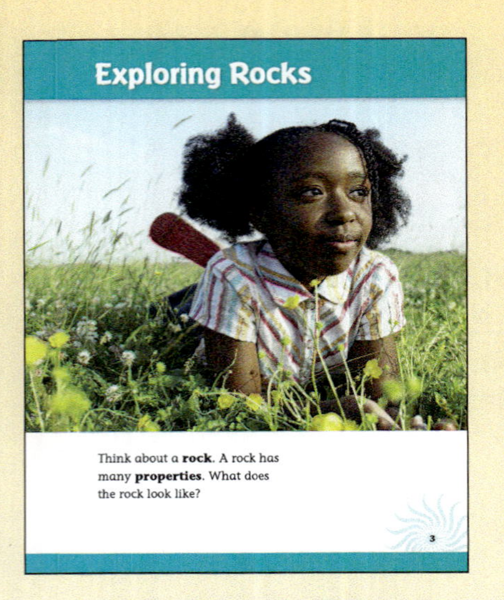

Exploring Rocks
Think about a **rock**. A rock has many **properties**. What does the rock look like?
3

ELA CONNECTION

These suggested strategies address the Common Core State Standards for ELA.

RI 4: Determine the meaning of words and phrases in the text.

RI 5: Know and use text features.

RI 6: Identify the main purpose of the text.

READING *in Science Resources*

15. Read "Exploring Rocks"

Give students a few minutes to look at and discuss the cover of *FOSS Science Resources: Pebbles, Sand, and Silt*. Have them examine the table of contents. If this is your class's first exposure to a table of contents, take a few minutes to discuss what a table of contents is and why it is used.

Point out the title of the article "Exploring Rocks" and have students turn and talk to a neighbor about what they think the main purpose of this text will be. Ask,

➤ *What do you think the author wants to explain or describe?* [Properties of rocks.]

Have students share what they have learned so far about the properties of rocks. Tell students this article will review what they have been learning and may have some new information.

Read the first page and pause to let students develop an image of a rock in their minds. Encourage them to compare their images with the possibilities suggested in the reading.

Read aloud or have students read the next page independently. Ask what they notice about the way the author describes the properties of rocks. [They are opposites.] Model how you might figure out what the word "dull" means if you know it is the opposite of "shiny." Continue reading the next two pages and ask students to think about and share with a partner any experiences they have had that support the idea that a rock can be as big as a mountain or as small as a grain of sand. What is the size of the rock they are imagining?

Next, have students look at the picture of the rocks in the river. Ask them to imagine how these rocks feel. [Smooth.] Why do you think they are smooth? Read the passage and ask students to take turns explaining to a partner how the rocks in a river become smooth. Have them compare the volcanic rocks to the river rocks they have explored. Why are the river rocks smoother? Show the video of the stream flowing over the rocks from page 7.

Read aloud or have students read the rest of the article independently.

16. Discuss the reading

Discuss the reading, using these questions as a guide.

➤ *Describe the rock you were thinking about.* [Students may want to draw a picture of their rocks and write or dictate a sentence describing them.]

Ask,

➤ *Now that we have read this article, what do you think the main purpose of the text is?* [Describe the properties of rocks; explain how rocks become smooth; students might also respond that the author wants the reader to explore rocks on their own.]

The following questions can be used to deepen students' understanding and may best be used after students have reread the article in small groups.

➤ *Where do we find rocks?* [Make a list of student responses.]

➤ *What are some properties of rocks?* [Size, shape, texture, color, pattern.]

Read *Everybody Needs a Rock* by Byrd Baylor. This is a good opportunity to compare and contrast how this idea is presented in two different texts.

SCIENCE AND ENGINEERING PRACTICES

Obtaining, evaluating, and communicating information

ELA CONNECTION

These suggested strategies address the Common Core State Standards for ELA.

RI 1: Ask and answer questions to demonstrate understanding.

RI 9: Compare and contrast two texts on the same topic.

WRAP-UP/WARM-UP

17. Share notebook entries

Conclude Part 4 or start Part 5 by having students share notebook entries. Ask students to open their science notebooks to the most recent entry. Read the focus question together.

➤ *What are the properties of schoolyard rocks?*

Ask students to work with a partner to

- share their answers to the focus question;
- describe their drawings.

Have students pair up with students from another group. Model how to give constructive feedback and ask students to critique each other's work. Have students tell their partner one thing they think is good about the entry and then one thing they could do to make it better. Allow time for students to add to or revise their entries based on the feedback from their partner.

To help students make detailed drawings, have them make another drawing of their rock on the next blank page using the ABCD (accurate, big, complete, and detailed) strategy. Have each table group leave their notebooks open to the page with their drawing and put their rocks all together in the center. Have the groups rotate to another group's table and see if they can match the rocks with the drawings. After students have rotated a few times, discuss what made it easier to identify the rocks in the drawing. Ask students how they might improve their drawings.

ELA CONNECTION

These suggested strategies address the Common Core State Standards for ELA.

SL 1: Participate in collaborative conversations.

W 5: Strengthen writing by revising and editing.

TEACHING NOTE

Refer to the Science Notebooks in Grades K–2 chapter for a description of the ABCD drawing strategy.

MATERIALS *for*

Part 5: *Sorting Activities*

For each pair of students

- 1 Bag of 20 large pebbles (from Part 3)
- 2 Sorting mats (See Step 3 of Getting Ready.)
- 2 Hand lenses
- 1 Piece of pink granite
- 2 *FOSS Science Resources: Pebbles, Sand, and Silt*
  - "Colorful Rocks"

For the class

- 1 Bag of 20 large pebbles (from Part 3)
- ❑ 1 Teacher master 3, *Sorting Mat Side One*
- ❑ 1 Teacher master 4, *Sorting Mat Side Two*
- 1 Document camera (optional) ★
- 1 Big book, *FOSS Science Resources: Pebbles, Sand, and Silt*
- 1 Computer with Internet access ★

For benchmark assessment

- ❑ • *Investigation 1 I-Check*
- ❑ • *Assessment Record*

★ Supplied by the teacher. ❑ Use the duplication master to make copies.

No. 3—Teacher Master

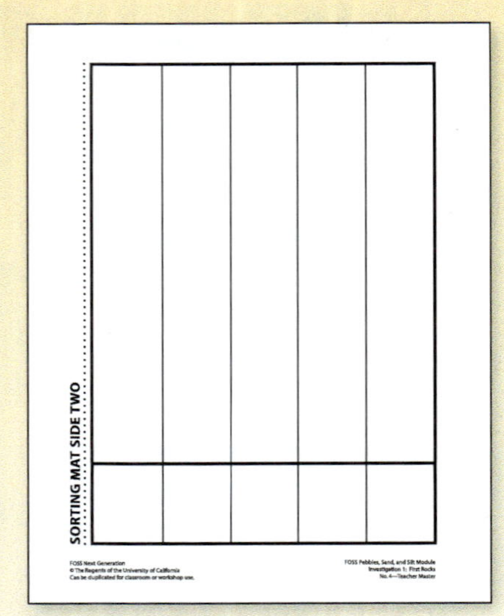

No. 4—Teacher Master

GETTING READY *for*

Part 5: *Sorting Activities*

1. Schedule the investigation

Plan one active investigation session for sorting activities and discussion. Plan a second session for the reading and a third short session for the I-Check.

If you feel that students have had enough rock-sorting experience for now, you may want to schedule this part as a review after you have completed some other activities.

2. Preview Part 5

Students use sorting mats to compare and sort the river rocks. The focus question is **How many ways can rocks be sorted?**

3. Prepare sorting mats

Each student will need a two-sided sorting mat made from teacher masters 3 and 4.

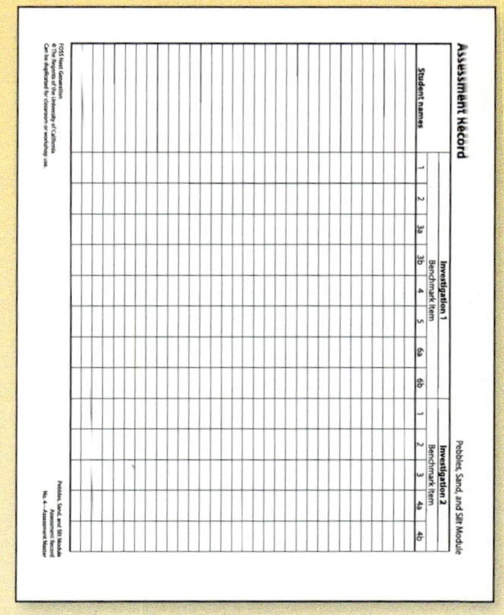
Assessment Record

Student names

Investigation 1 Benchmark Item: 1, 2, 3a, 3b, 4, 5, 6a, 6b

Investigation 2 Benchmark Item: 1, 2, 3, 4a, 4b

Pebbles, Sand, and Silt Module

No. 3—Assessment Master

4. Plan to use a document camera (optional)

A document camera can be used to introduce the sorting activities described in Step 3 of Guiding the Investigation.

5. Plan for online activities

Preview the online activities "Rock Sorting" and "Property Chain." Students have more opportunities to sort and group rocks by property.

The link to these activities for teachers is in the Resources by Investigation on FOSSweb.

6. Plan to read *Science Resources*: "Colorful Rocks"

Plan to read "Colorful Rocks" during a reading period after completing this part.

7. Plan assessment: I-Check

Plan to give *Investigation 1 I-Check* at the end of the investigation. Read the items aloud to the whole class, and have students answer independently. Review students' responses using the What to Look For information in the Assessment chapter. Use assessment master 3, *Assessment Record,* to record students' responses.

GUIDING *the Investigation*

Part 5: *Sorting Activities*

FOCUS QUESTION

How many ways can rocks be sorted?

Materials for Steps 3–4

- *Bags of large pebbles*
- *Sorting mats*

1. Review rock properties

Call students to the rug. Review some of the properties of the schoolyard rocks. Tell students,

Geologists sometimes see things they hadn't noticed before when they observe rocks for a second time. Let's take another look at our river rocks to see what new properties we might observe. We will do that by sorting rocks.

2. Focus question: How many ways can rocks be sorted?

Ask the focus question as you write it on the board.

➤ *How many ways can rocks be sorted?*

3. Introduce sorting activities

Show students a copy of the two-sided sorting mat or project the sheets. Point out the circles on one side and the lines on the other side of the mat. Tell students that you have some sorting activities to show them. Use a sorting mat and your set of river rocks to demonstrate. You can select a few of the activities to show them first and then introduce other activities to the class or to groups later in the session. Students might invent their own activities as they work.

- Mystery Pairs—Using *Sorting Mat Side One*, have one student choose two rocks and place the pair in one of the circles. The second student tries to figure out the rule the first student used to pair the rocks. Now it's the second student's turn to choose a mystery pair and let the first student do the guessing.
- Property Chain—Using *Sorting Mat Side Two* held horizontally, one student chooses a rock and places it in the top row. A second student chooses a rock that shares one property of the first student's rock (e.g., rough, smooth), and places it to the right of the first rock (in the same row). The next rock placed must share at least one property of the previously placed rock. Continue taking turns and placing rocks across the mat until the top two rows are filled. Challenge students to name the property they are thinking of as they place each rock on the mat.

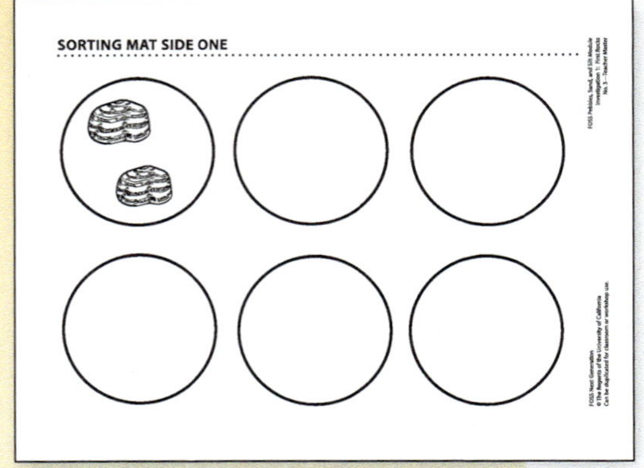

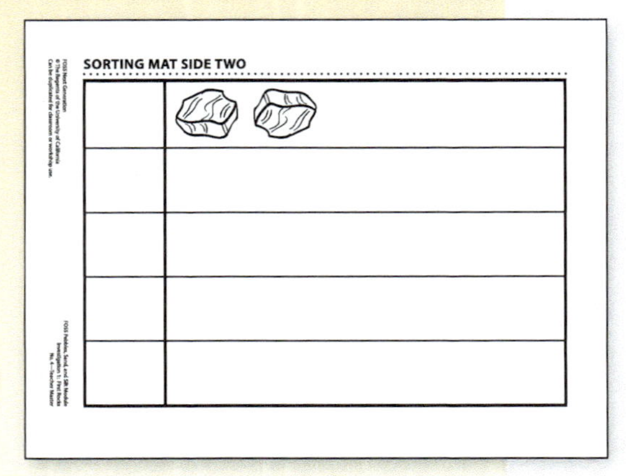

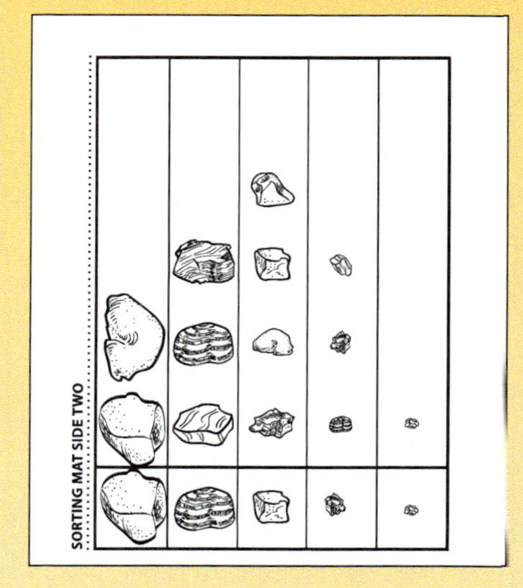

- Stand Up and Be Counted—Using *Sorting Mat Side Two* held vertically, small squares at the bottom, sort the rocks by one property (such as size or color). If each column represents a subset of the property, there can be up to five groupings. Place like rocks in a column, one above the other. Encourage students to use this to discover which attribute is most common or the total number of a certain kind of rock.
- Shape Sorter—Use *Sorting Mat Side Two* held vertically, but with the small squares at the top. Have students observe all their rocks and identify up to five geometric shapes they see. Using a crayon, have them draw a represented shape in each column head (triangle, circle, square, rectangle). Sort the rocks and place them in the appropriate columns. Make comparisons about the total number of each shape.

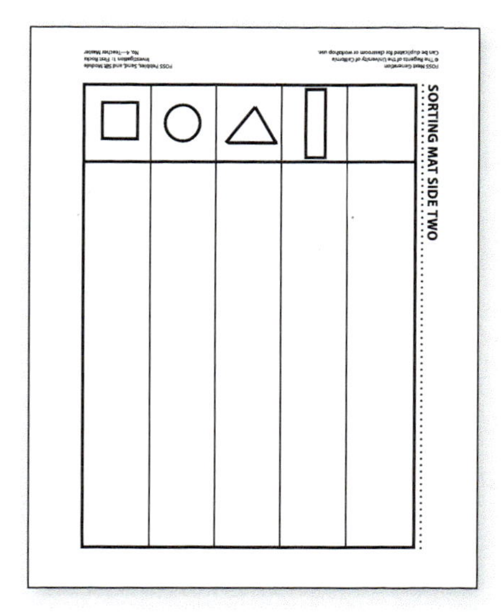

- Line Up, in Order—Using all the rocks, have students choose a property (size or texture), and place all the rocks in order from smallest to largest or smoothest to roughest.

4. Clean up

Have Getters return the rock sets and mats to the materials station.

TEACHING NOTE

See the ***Home/School Connection*** *for Investigation 1 at the end of the Interdisciplinary Extensions section. This is a good time to send it home with students.*

Colorful Rocks

What are these colorful objects?

They are **minerals**. There are many different kinds of minerals. Minerals come in lots of different colors.

11

Materials for Step 6–8

- *Hand lenses*
- *Pieces of granite*

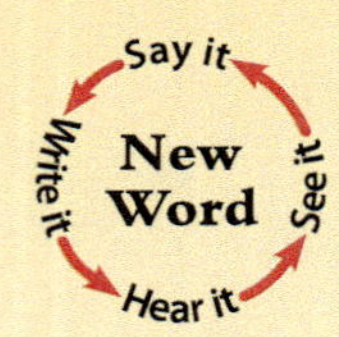

ELA CONNECTION

These suggested strategies address the Common Core State Standards for ELA.

RI 1: Ask and answer questions to demonstrate understanding.

RI 4: Determine the meaning of words and phrases in the text.

RI 7: Explain how images contribute to and clarify text.

W 7: Record science observations.

SCIENCE AND ENGINEERING PRACTICES

Obtaining, evaluating, and communicating information

READING *in Science Resources*

5. Read "Colorful Rocks"

Ask students what makes rocks so colorful. Discuss students' ideas as a class.

Distribute a sample of pink granite to each pair of students and ask them what they observe. Students should notice that granite is made of different colors. Ask students what questions they have about the granite. Tell them that the article "Colorful Rocks" might answer some of their questions about why rocks are different colors. Let students look at and discuss the pictures with a neighbor and then read aloud or have students read the article independently.

6. Discuss the reading

Discuss the reading, using these questions as a guide.

➤ *What makes rocks colorful?* [Minerals.]

➤ *How many different kinds of* ***minerals*** *can you see in* ***granite****? What color are they?* [Pink, gray, and black]

Add *mineral* and *granite* to the word wall.

Ask students to explain how the images help them understand the text. Students should respond that the enlarged photograph makes it easier to see the minerals in the granite. Or that the photograph of several colored minerals shows how different they are in color.

7. Observe granite

Return students attention to the samples of pink granite. Provide hand lenses and ask students to identify the different minerals by color in the granite. Have them draw the granite in their notebooks as they see it through the hand lens. They should add the name of the rock and a detailed description.

If a class of older students is doing the **Soils, Rocks, and Landforms Module**, invite some students from that class to come to your classroom to share and discuss the granite samples.

8. View online activities: "Rock Sorting" and "Property Chain"

In small groups or as individuals, have students engage with the online activities focusing on rock properties and grouping. The link to these activities for teachers is in the Resources by Investigation on FOSSweb, and for students, in the Online Activities.

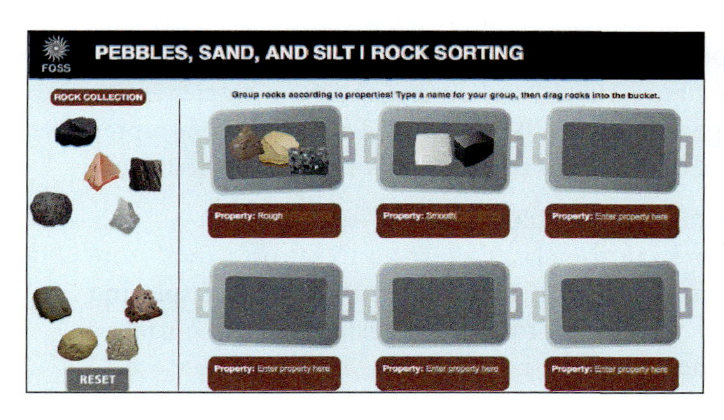

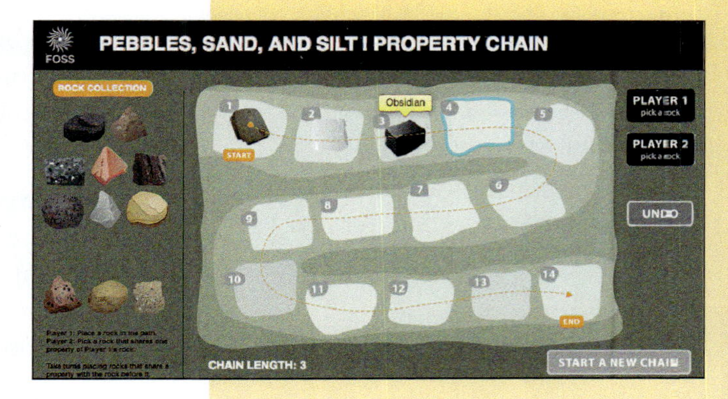

9. Review properties and changes

Conduct a brief discussion with students to revisit the properties of rocks and the guiding question of how rocks change.

➤ *What are properties of rocks and how do they change?*

BREAKPOINT

10. Assess progress: I-Check

When students have completed the investigation, give them *Investigation 1 I-Check.*

Review student responses. Use the What to Look For information in the Assessment chapter for guidance. Note concepts that you might want to revisit with students, using the next-step suggestions.

The students' experiences in this investigation contribute to their understanding that some events on Earth happen quickly and others occur over a very long period of time, and that matter can be described by its observable properties.

DISCIPLINARY CORE IDEAS

ESS1.C: The history of planet Earth

PS1.A: Structure and properties of matter

TEACHING NOTE

Refer to the teacher resources on FOSSweb for a list of appropriate trade books that relate to this module.

TEACHING NOTE

Encourage students to use the Science and Engineering Careers Database on FOSSweb.

INTERDISCIPLINARY EXTENSIONS

Language Extensions

- **Make a rock record book**
  As the class rock collection grows, it can become the motivator for a class book. Have students complete a rock record for each rock they contribute; these can become the page entries in the book. Encourage students to be as detailed as they can in representation and description. Have students take photos of their rocks.

- **Set up a rock store**
  Use the rocks in the class collection to set up a rock store. Have students use the sorting mats for arranging and displaying their rocks. Ask them to write descriptions of the rocks as advertisements. Invite parents in to "shop" at the store.

- **Make stone soup**
  Read *Stone Soup* by Ann McGovern to the class. Make stone soup as a class project.

- **Write about magic pebbles**
  Read *Sylvester and the Magic Pebble* by William Steig to the class. Have students write a short story about their own imaginary magic pebble.

- **Read about special rocks**
  Read *Everybody Needs a Rock* by Byrd Baylor. The book suggests ten rules for finding a special rock. Ask students if they agree with the rules or if they would change them. Discuss how or why students might change the rules.

- **Create a property map**
  As a whole class or with partners, have students create a property map for the word *rocks.* Categorize the terms according to shape, size, color, and texture (see example). Students may use vocabulary from the word wall, the article "Exploring Rocks," or their own experiences.

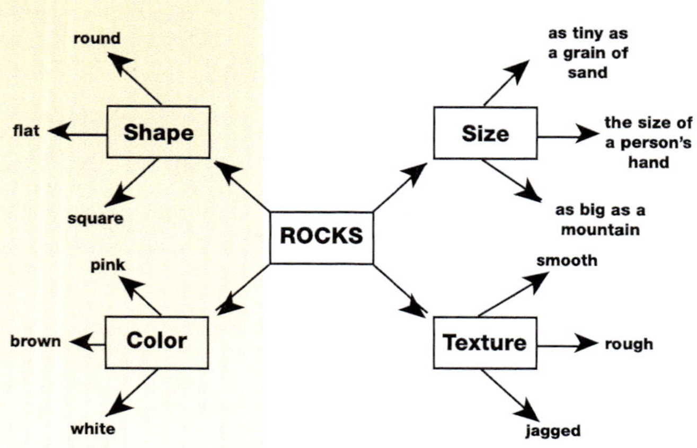

Math Extensions

- **Math problem A**

  In this problem, pairs of students will share the bag of 20 river rocks from Part 5. Each student selects rocks and sorts them on the sheet by any two properties. They use the rock groups to create number sentences.

  Notes on the problem. This problem provides practice with combinations of numbers to 20. The sheet is designed so that you can have the entire class work with the same number of rocks *or* you can scale the problem based on individual students' skills and abilities and assign different numbers of rocks.

  Students sort a given number of rocks into two groups. This is an opportunity to sort by one attribute that is present or not present. For example, red rocks and not-red rocks. This creates two clear groups and is an example of bifurcation or separation of one group into two groups. After the rocks are sorted, students label the sort, count the number in each group, and generate a number sentence (equation) for the total number of rocks.

  Next, students gather the same rocks and sort them a second time into two groups using a different property, such as size or texture. Again, they label the sort, count each group, and record an equation for the sort.

  Before students work independently, show how to sort rocks into two groups and model how to record an equation for the sort.

- **Math problem B**

  This problem asks students to create a story problem to go with the illustrated rocks.

  Notes on the problem. To introduce the problem, provide a context for students to write a story about rocks involving addition or subtraction. Tell them that a girl went on a rock hunt and found some rocks. Show a collection of rocks or a picture of rocks (but not the one used in Math Extension B). Count the rocks and make observations about them.

  Tell students that you want to write an addition story about the rocks. As a class, generate a list of words (for later reference) that could be used in the story to tell about the rock collection. Create a story as a class to model the process of writing a problem.

  Show students the sheet for problem B with the rocks a boy collected. Tell students that they will write their own number story that goes with the picture of the boy's rock collection.

TEACHING NOTE

Examples of number sentences

4 dark rocks and 6 light rocks total 10 rocks.

3 big rocks and 6 small rocks total 9 rocks.

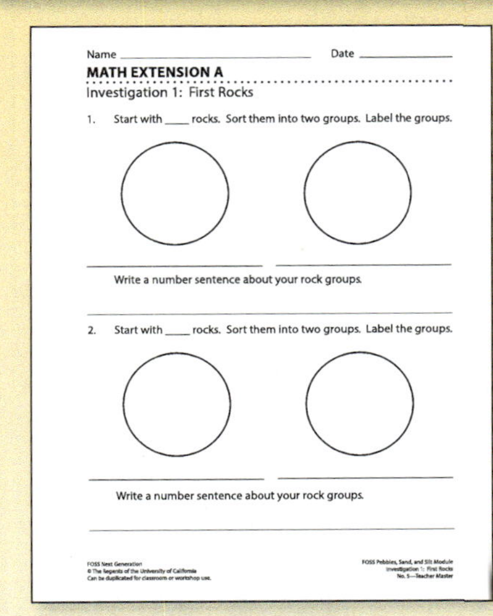

Name ______ Date ______

MATH EXTENSION A

Investigation 1: First Rocks

1. Start with ___ rocks. Sort them into two groups. Label the groups.

Write a number sentence about your rock groups.

2. Start with ___ rocks. Sort them into two groups. Label the groups.

Write a number sentence about your rock groups.

No. 5—Teacher Master

Name ______ Date ______

MATH EXTENSION B

Investigation 1: First Rocks

Write a story problem to go with this rock collection.

No. 6—Teacher Master

- **Graph rock sorts**

Sort rocks to create a bar graph. *Sorting Mat Side Two* works well to help students organize the graph. Before students work independently, model how to sort and create a graph of rocks.

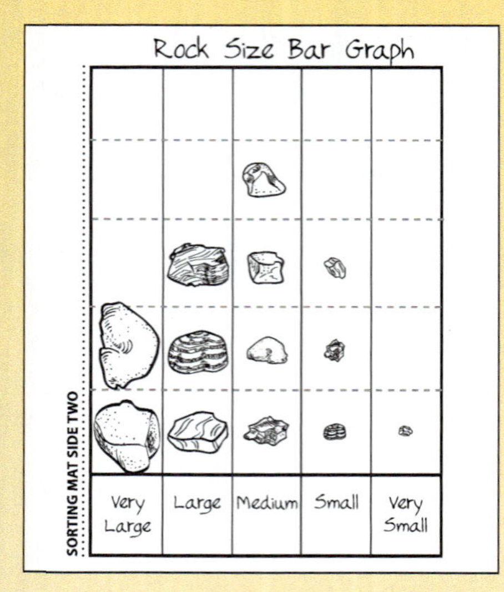

Show your collection of rocks and have students suggest ways to sort the rocks. Select one method of sorting, such as size, and sort. Tell students that you want to compare the number of rocks in each size group and that you have a tool to help. It is a graphing grid. Show the sorting mat, holding it vertically. In the small boxes at the bottom of the sheet, record the properties used to sort, such as the sizes of the rocks. Then place the rocks in the appropriate column. Be aware that as the columns are created, the rocks should line up horizontally *regardless* of their size. That ensures an accurate comparison of the rocks in each group.

When the concrete graph is complete, have students make statements about the graph. Use statements such as, "In this collection of rocks, there were more medium rocks than any other size," or "We had the fewest very small rocks." Steer students away from seeing the graph as a competition in which one type of rock "wins."

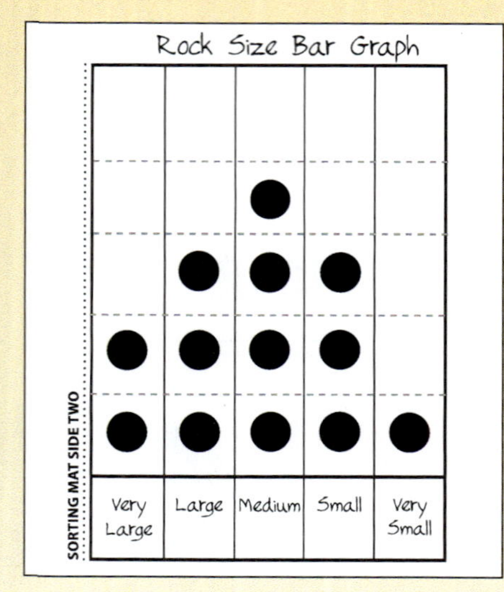

Tell students that you want to save this information and that you are going to record it, using circles to represent the real rocks. On a blank *Sorting Mat Side Two,* record the sizes and draw a small circle that represents each rock on the concrete graph. Have students compare the concrete graph to the representational graph. Check that the data is the same. Add a title to the graph.

Have students work with a partner to sort and concretely graph a collection of rocks. Then have each student create a representational graph of the data. Have students take a tour of the graphs to see both the concrete and the representational graphs.

Art Extensions

- **Build rock towers**
  Have students collect a number of different-sized rocks. Find a location in the schoolyard for students to assemble a tower of rocks. Students must work with the natural shape of the rocks to balance them in one tower. Have students share the properties that helped each rock balance.

- **Make rock people or pets**
  Have students collect different-sized river rocks that they can use to create rock pets. Let them paint the rocks with poster paint, glue the rocks together, and apply wiggly eyes and yarn or felt for hair. One FOSS teacher suggested calling the rock people "Reading Rocks" and asked students to read to their rocks every night.

- **Assemble a rock aquarium**
  Get a clear container, such as a recycled plastic bottled-water jug (5 gallons) or a plastic aquarium, and fill it with water. Let students put their favorite rocks into the water for an aesthetic display.

- **Make rock checkers**
  Have students put together two sets of checkers made from rocks of different colors. Let them make a checkerboard and use the rocks to play a game of rock checkers.

- **Find out about rock games**
  A number of games use rocks as playing pieces. Find out about these games (e.g., mankala, an African stone game; hopscotch; and marbles) and introduce them to students.

Science Extensions

- **Find your rock**
  Have students carefully observe their favorite river rocks and then put them in a class rock pile. Ask students to find them again when the pile is all mixed up.

- **Start personal rock collections**
  People who collect rocks are known colloquially as rock hounds. Encourage students to make their own rock collections at home. Let students share their collections or ask them to bring in pictures of their collections to post on a bulletin board for rock hounds.

TEACHING NOTE

Review the online activities for students on FOSSweb for module-specific science extensions.

HOME/SCHOOL CONNECTION
Investigation 1: First Rocks

Invent a game that uses different kinds of rocks. It should use the properties of rocks you have.

Here are some examples. The goal of the game could be to put together similar-looking rocks, like the game rummy. Or the goal could be finding one rock among many. Or the goal could be to find ways that rocks are the same, as in the games dominoes or crazy eights. It could also be a brand new game that you invent.

Have a family member help you write the directions for the game so you can share it in class.

No. 7—Teacher Master

Home/School Connection

Students and family members invent a game that uses rocks. Games should use the properties of rocks students have at home. Students will need to enlist the aid of a family member to help write their directions to share in class.

Print or make copies of teacher master 7, *Home/School Connection* for Investigation 1, and send it home with students during Part 5.

As another home/school connection, ask students to talk with their families about cultural connections to the use of rocks for games, in addition to other uses such as art, music, tools, and spiritual traditions.

Investigation 2: River Rocks

INVESTIGATION 2 — *River Rocks*

Guiding question for phenomenon:
How are small pieces of rock made and moved to change landforms?

PURPOSE

Students investigate and sort river rock aggregates as a phenomenon. Through observation, discussion, and media, students experience how a single material, rock, can be present in different particle sizes. These particles, a result of weathering, are moved by wind and water to change landforms.

Content

- Rocks are earth materials and can be described by the property of size—clay, silt, sand, gravel, pebbles, cobbles, and boulders.
- Weathering, caused by wind or water, causes larger rocks to break into smaller rocks.
- Some Earth events happen very quickly (volcanic eruptions, floods); others occur very slowly over a long period of time (weathering and movement of rock).

Practices

- Separate an earth material mixture into rock particles of various sizes using screens and water.
- Compare properties (size, shape) of different landforms.

Science and Engineering Practices

- Developing and using models
- Planning and carrying out investigations
- Analyzing and interpreting data
- Using mathematics and computational thinking
- Constructing explanations
- Engaging in argument from evidence
- Obtaining, evaluating, and communicating information

Disciplinary Core Ideas

ESS1: What is the universe, and what is Earth's place in it?

ESS1.C: The history of planet Earth

ESS2: How and why is Earth constantly changing?

ESS2.A: Earth materials and systems

ESS2.B: Plate tectonics and large-scale system interactions

ESS2.C: The roles of water in Earth's surface processes

PS1: How can one explain the structure, properties, and interactions of matter?

PS1.A: Structure and properties of matter

Crosscutting Concepts

- Patterns
- Cause and effect
- Scale, proportion, and quantity
- Stability and change

INVESTIGATION 2 — *River Rocks*

	Investigation Summary	Time	Focus Question for Phenomenon, Practices
PART 1	**Screening River Rocks** Students separate a river-rock mixture, using a set of three screens. They discover five sizes of materials: large pebbles, small pebbles, large gravel, small gravel, and sand.	**Active Inv.** 1–2 Sessions *	**How can rocks be separated by size?** **Practices** Planning and carrying out investigations Analyzing and interpreting data Using mathematics and computational thinking Constructing explanations
PART 2	**River Rocks by Size** Students use squares of three sizes as a tool (instead of screens) to seriate rock particles into sand, gravel, and pebbles.	**Active Inv.** 1 Session **Reading** 1 Session	**How else can rocks be sorted by size?** **Practices** Developing and using models Analyzing and interpreting data Constructing explanations Obtaining, evaluating, and communicating information
PART 3	**Sand and Silt** Students take a close look at sand and separate sand particles from silt particles, which are smaller than the sand, by mixing the sand with water and allowing the particles to settle. They observe that the sand settles to the bottom and the silt forms a layer on top of the sand.	**Active Inv.** 2 Sessions	**Is there an earth material smaller than sand?** **Practices** Planning and carrying out investigations Analyzing and interpreting data Engaging in argument from evidence
PART 4	**Exploring Clay and Landforms** Students investigate the properties of the smallest rock particles, clay. They read about and view a video about ways that wind and water move and shape the land. Students compare the time it takes to change the surface of the land.	**Active Inv.** 2 Sessions **Reading** 2 Sessions **Assessment** 1 Session	**What earth material is smaller than silt?** **How does water and wind change landforms?** **Practices** Developing and using models Planning and carrying out investigations Analyzing and interpreting data Constructing explanations Obtaining, evaluating, and communicating information

* A class session is 45–50 minutes.

Content Related to DCIs	Writing/Reading	Assessment
• Rocks can be described by the property of size. • Screens can be used to sort the sizes of earth materials. • Rock sizes include sand, small gravel, large gravel, small pebbles, and large pebbles. • Rocks are earth materials.	**Science Notebook Entry** *Screening River Rocks* *Rock Graph* (optional)	**Embedded Assessment** Performance assessment
• Rocks can be categorized visually by size. • Rock sizes include sand, small gravel, large gravel, small pebbles, and large pebbles. • Rocks larger than pebbles are cobbles. • Rocks larger than cobbles are boulders. • Smaller rocks result from the weathering of larger rocks.	**Science Notebook Entry** *River Rocks by Size* *The Story of Sand* (optional) **Science Resources Book** "The Story of Sand"	**Embedded Assessment** Science notebook entry
• Sand often contains smaller particles, called silt. • Water can be used to sort the sizes of earth materials.	**Science Notebook Entry** *Sand and Water Drawing*	**Embedded Assessment** Science notebook entry
• Clay particles are very small, even smaller than silt. • Weathering, caused by wind or water, causes larger rocks to break into smaller rocks. • Some Earth events happen very quickly (volcanic eruptions, floods); others occur very slowly over a long period of time (weathering of rock).	**Science Notebook Entry** *Clay and Water Drawing* *Rocks in Bottle Drawing* **Science Resources Book** "Rocks Move" "Landforms" **Video** *All about Land Formations*	**Embedded Assessment** Science notebook entry **Benchmark Assessment** *Investigation 2 I-Check* **NGSS Performance Expectations addressed in this investigation** 2-ESS1-1 2-ESS2-1 2-ESS2-2 (foundational) 2-ESS2-3 2-PS1-1

BACKGROUND *for the Teacher*

The solid, liquid, and gaseous materials that make up Earth and its atmosphere are known collectively as earth materials. Far and away the most common solid material at Earth's surface is rock—the phenomenon of this investigation.

As soon as a mass of rock cools from molten magma or lava, or is raised up from the bottom of the sea, weathering processes on Earth's surface begin to break it down. Cracks in rock fill with rainwater. If the temperature falls low enough, the water freezes. Water expands as it freezes, and the crack widens. After many seasons of freezing and thawing, a chunk of rock might finally break off and fall, hitting other rocks as it tumbles. This process is one example of physical weathering.

Other types of physical weathering can reduce a large mass of rock to a pile of small **particles**, the size of **sand** or **silt**. Drought, salt crystallization, and growing plants can force rocks apart.

Water and the materials dissolved in it can cause the chemical decomposition of the minerals that make up rocks. Chemical weathering can weaken rocks and cause them to fall apart.

It is important to realize that the immense monolithic structures in the Rocky Mountains, the **boulders** over which rivers fall, the **cobbles** that result from weathering, and the minute particles of matter used to make a ceramic coffee mug are made of similar materials. It is just their geological history that contributes to their different appearance today.

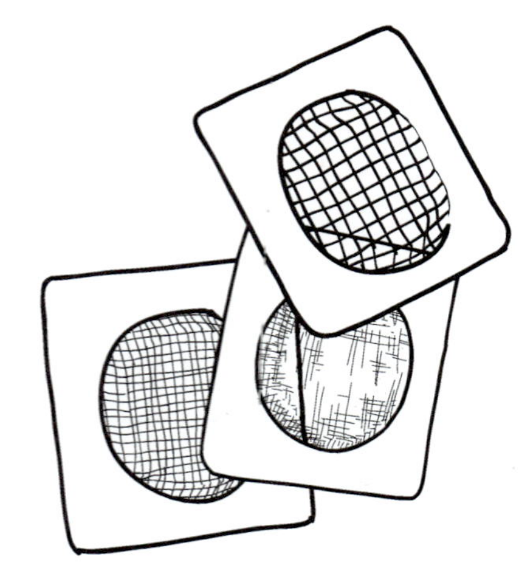

Clay	Less than 0.004 mm
Silt	0.004–0.062 mm
Sand	0.062–2 mm
Gravel	2–4 mm
Pebble	4–64 mm
Cobble	64–256 mm
Boulder	More than 256 mm

How Can Rocks Be Separated by Size?

Rock that has broken away from the continuous **layer** of bedrock is found in particles that range from the minuscule to the gigantic. Geologists have devised several schemes to classify rock particles by size. The Wentworth scale is the one most frequently used. For FOSS investigations, these size ranges (diameter in millimeters) are used to describe rock particles.

Students sort and group river rocks based on their sizes. Most students can do this mechanically by picking out rock bits by hand. But in this investigation, students will use **screens** of different sizes to **separate** dry **mixtures** of river aggregate into sand, **gravel**, and **pebbles**.

How Else Can Rocks Be Sorted by Size?

Another way to separate rocks into sizes is to use a representation of a screen on a piece of paper. The tool is slightly different, but the process and the end result are the same. The particles can be seriated from smallest to largest.

Students can simulate some of the processes that separate rock particles in the natural world by **shaking**, blowing, screening, and mixing earth materials with water. Each separation provides students with a little more information about the properties and behavior of earth materials.

Is There an Earth Material Smaller than Sand?

Rocks are often found in mixtures that include many sizes. The size and distribution of rock particles give geologists clues about earth-shaping processes. Currents of water or air can sort rock particles by size. The stronger or faster a stream is flowing or wind is blowing, the larger the particle size it can carry. Floods can move huge boulders. A swiftly moving stream can carry gravel. It doesn't take as much energy to move sand, and even less for the smaller particles of silt and **clay**. Silt and clay can stay suspended in water in relatively slow-moving streams. Clay particles eventually **sink** and **settle** out of water but this happens very slowly. Deposits of clay are usually associated with lakes, marshes, or deep marine environments, where the water is usually still and quiet.

What Earth Material Is Smaller than Silt?

It is difficult to determine the difference between silt and clay particles with the naked eye, but some useful clues help with their identification. With a hand lens you can discern individual particles of silt, but clay particles are too small to see with a lens. Silt feels somewhat rough when you rub it between your fingers. Clay feels smooth or even slimy. Clay holds its shape when squeezed together in the hand and dries into a hard, compact lump overnight. These properties make clay useful to potters and brick makers. Silt settles out of water faster than clay and forms a layer below clay. The settling rate of clay helps manufacturers separate it from silt and into several grades.

How Does Water and Wind Change Landforms?

Some events change the surface of Earth quickly. Eruptions of volcanoes due to pressures that build up in the interior of Earth can cause changes relatively quickly. Strong, moving water during storms and floods is a powerful force and can change Earth's surface in a hurry. But some of the biggest changes in Earth's surface are caused when moving water has a long time to work through the processes of weathering and **erosion**. The biggest changes occur over a long period of time. Not even mountains and **plateaus** can stand up to moving water wearing away at them for thousands of years. Wind can also cause erosion—movement of particles from one place to another. Many land formations result—**valleys**, **canyons**, **buttes**, **mesas**, **sand dunes, deltas**, **beaches**, and **plains**. We can **model** these landforms by making physical representations, drawings, or diagrams.

TEACHING CHILDREN *about River Rocks*

Developing Disciplinary Core Ideas (DCI)

NGSS Foundation Box for DCI

ESS1.C: The history of planet Earth
- Some events happen very quickly; others occur very slowly over a time period much longer than one can observe. (2-ESS1-1)

ESS2.A: Earth materials and systems
- Wind and water can change the shape of the land. (2-ESS2-1)

ESS2.B: Plate tectonics and large-scale system interactions
- Maps show where things are located. One can map the shapes and kinds of land and water in any area. (2-ESS2-2, foundational)

ESS2.C: The roles of water in Earth's surface processes
- Water is found in the ocean, rivers, lakes, and ponds. Water exists as solid ice and in liquid form. (2-ESS2-3)

PS1.A: Structure and properties of matter
- Different kinds of matter exist and many of them can be either solid or liquid, depending on temperature. Matter can be described and classified by its observable properties. (2-PS1-1)

Most children love to play in the sand and mud. It is a multisensory experience people of all ages can relate to. Good educational experiences build on students' innate enthusiasm for messing around in the sand and soil, and allow students to develop and express their understanding of the natural world.

Observing, communicating, and comparing are the thinking processes most powerful for primary students. *River Rocks* offers students opportunities to use these processes.

Many primary students need multiple experiences using screens to separate a rock mixture. Setting up a screening center in a corner of the room or even on the playground can provide this extra experience. After several opportunities to work informally with the screens, these students will be ready for more guided use of the screens to separate the materials and keep them separated for further observation.

In this investigation, students also shake a mixture in water to observe the layering of different-sized particles. When students have experiences using the tools to separate a variety of earth materials, they can compare the results and come to a better understanding of the properties of the materials that come from Earth.

Primary students are able to comprehend that large pebbles, gravel, and even sand are all essentially the same material, just broken into larger or smaller pieces. Students can verify this by observation of these macroscopic materials—they look the same. A simple hand lens will help students recognize sand grains as tiny rocks. However, the microscopic sizes of rock (silt and clay) are a little harder for students to conceptualize as bits of rock. Silt feels rough, but individual pieces can't be seen by the unassisted eye. Wet clay feels smooth or slick and doesn't have discernible particles.

Many students may not appreciate the rock size continuum, and that's all right. The idea that sand found on a beach was weathered from a massive rock (a mountain) a long time ago, moved by wind and water to its present location, is not intuitive. It requires students to think about what might have happened to transform and deliver a tiny piece of rock to its present location. The important thing is for students to experience the properties of these important earth materials in an appropriate context,

so that when later, in grade four, they are confronted with more complex notions of the processes of erosion and deposition of the mineral part of Earth, they will be better grounded and better able to integrate their primary experiences into more advanced concepts. Simple notions of the affect of wind and water on earth materials is the purpose of these first-hand investigations with rocks.

As students work their way through the series of rock sizes, they exercise a rich vocabulary to describe particle sizes and earth-material properties. These vocabulary words should be introduced after students have had concrete experience with the materials and phenomena.

The activities and readings students experience in this investigation contribute to the disciplinary core ideas **ESS1.C, The history of planet Earth; ESS2.A, Earth materials and systems; ESS2.B, Plate tectonics and large-scale system interactions; ESS2.C, The roles of water in Earth's surface processes;** and **PS1.A, Structure and properties of matter.**

Engaging in Science and Engineering Practices (SEP)

In this investigation, students engage in these practices.

- **Developing and using models** to represent how earth materials break down into smaller pieces (weathering). Students use models to explain how rock pieces interact with water and how to describe landforms. They compare the model to the actual object.
- **Planning and carrying out investigations** with rocks to separate them into sizes using a variety of tools (screens, water, settling). Students collect data from firsthand investigations to make comparisons of rocks.
- **Analyzing and interpreting data** from investigations with rocks of different sizes to explore their properties and their behavior in water.
- **Using mathematics and computational thinking** by measuring and comparing rock particle size to sort rocks and describe categories (pebble, gravel, sand).
- **Constructing explanations** on the effects of wind and water on land formations.
- **Engaging in argument from evidence** about the results of investigations to find a rock particle smaller than sand.
- **Obtaining, evaluating, and communicating information** about earth materials and processes that change the shape of the surface of Earth by text, photos, diagrams, models, and videos.

NGSS Foundation Box for SEP

- **Distinguish between a model** and the actual object, process, and/or events the model represents.
- **Develop and/or use a model** to represent amounts, relationships, relative scales, and /or patterns in the natural world.
- **Develop or use a simple model based on evidence** to present a proposed object or tool.
- **Plan and conduct an investigation** collaboratively to produce data to serve as the basis for evidence to answer a question.
- **Make predictions** based on prior experiences.
- **Make observations** to collect data that can be used to make comparisons.
- **Record information** (observations, thoughts, and ideas).
- **Use and share pictures, drawings,** and/or writings of observations.
- **Use observations** to describe patterns and/or relationships in the natural world in order to answer scientific questions.
- **Describe, measure, and/or compare quantitative** attributes of different objects and display the data using simple graphs.
- **Analyze data** from tests of a tool to determine if it works as intended.
- **Make observations** to construct an evidence-based account for natural phenomena.
- **Construct an argument** with evidence to support a claim.
- **Read grade-appropriate text** and/or use media to obtain scientific information to determine patterns about the natural/designed world(s).
- **Obtain information** using various texts, text features, and other media that will be useful in answering a scientific question.

Exposing Crosscutting Concepts (CC)

In this investigation, the focus is on these crosscutting concepts.

- **Patterns.** One way that the earth material rock can be described and categorized is by particle size. Each particle size has different properties.
- **Cause and effect.** Weathering causes large rocks to break into small rocks. The effect is rocks of different sizes that can be sorted and described. When rocks of different sizes are put in water, the larger particles go to the bottom and the rest form layers in an observable pattern.
- **Scale, proportion, and quantity.** Rocks can be compared by their relative size; processes can be compared by the relative amount of time it takes for the process to occur.
- **Stability and change.** Earth changing processes can be compared. Some happen slowly (weathering and erosion) and some rapidly (volcanic eruptions, floods).

NGSS Foundation Box for CC

- **Patterns:** Patterns in the natural and human designed world can be observed, used to describe phenomena, and used as evidence.
- **Cause and effect:** Events have causes that generate observable patterns; simple tests can be designed to gather evidence to support or refute student ideas about causes.
- **Scale, proportion, and quantity:** Relative scales allow objects and events to be compared and described (bigger and smaller; hotter and colder; faster and slower).
- **Stability and change:** Some things stay the same while other things change. Things may change slowly or rapidly.

Connections to the Nature of Science

- **Scientific investigations use a variety of methods.** Scientific investigations begin with a question. Scientists use different ways to study the world.
- **Scientific knowledge is based on empirical evidence.** Scientists look for patterns and order when making observations about the natural world.
- **Science is a way of knowing.** Science knowledge informs us about the world.
- **Science addresses questions about the natural and material world.** Scientists study the natural and materials world.

Beach
Boulder
Butte
Canyon
Clay
Cobble
Delta
Erosion
Gravel
Layer
Mesa
Mixture
Model
Particle
Pebble
Plain
Plateau
Sand
Sand dune
Screen
Separate
Settle
Shake
Silt
Sink
Valley

Conceptual Flow

The earth material, rock, is found in different sizes, from huge, rough mountains to tiny, fine clay. That's the phenomenon students investigate. The guiding question for this investigation is how are small pieces of rock made and moved to change landforms?

The **conceptual flow** for this second investigation starts with students observing a mixture of earth materials referred to as river rocks. In Part 1, students are challenged to **separate** a river-rock **mixture** by **size** of the **particles**. They figure out how to use **screens** with three different sizes of mesh to separate the mixture. This is the first introduction to this geologist's tool that will also come into play when studying soil in Investigation 4. The separation results in five sizes of materials—large **pebbles**, small pebbles, large **gravel**, small gravel, and **sand**. Students are also introduced to a bar graph that visually displays and easily compares the number of particles of two sizes in a mixture.

In Part 2, students separate the same river-rock mixture using representational screens on paper. Three boxes, each representing the largest screen that a particle could go through, take the place of the screens. Students seriate the five particle sizes in their notebooks and make a permanent record. This reinforces the concept that these different earth materials are all rock of different sizes on a continuum from small particle to large. The focus then turns to the smallest particle, sand, and students read and discuss the story of sand formation through weathering of larger rocks, **boulders** and **cobbles**.

In Part 3, students investigate sand, using more techniques of the geologist—**shaking dry particles** to sort them by size and taking a sample and **mixing with water** to observe how the **particles** sink and settle out over time. Students find that in with the sand was a **smaller particle size, called silt**, which **forms a layer** on top of the sand when mixed with water. They make close visual and tactile observations of sand and silt

In Part 4, students explore **clay**, an earth material with a particle size even smaller than silt. They work with the clay dry and experience the small particles on their hands and desks. They allow some clay to dry while they wet another sample and agitate to observe what happens. They record their observations and compare the properties of clay to silt and sand. Then they view photographs of environments where these small earth materials have been **moved** by **wind** and **water** in mudflats, sand dunes, and hillsides. Through video, students observe the slow landform changes that result from weathering and **erosion**.

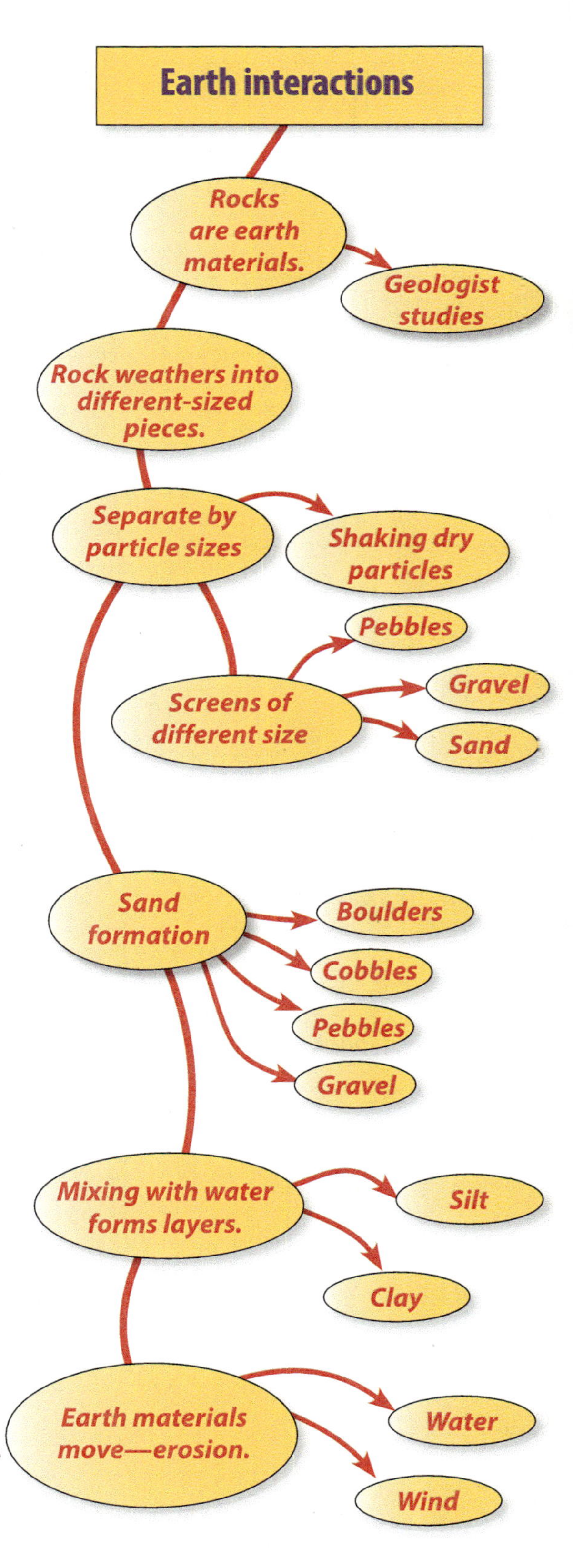

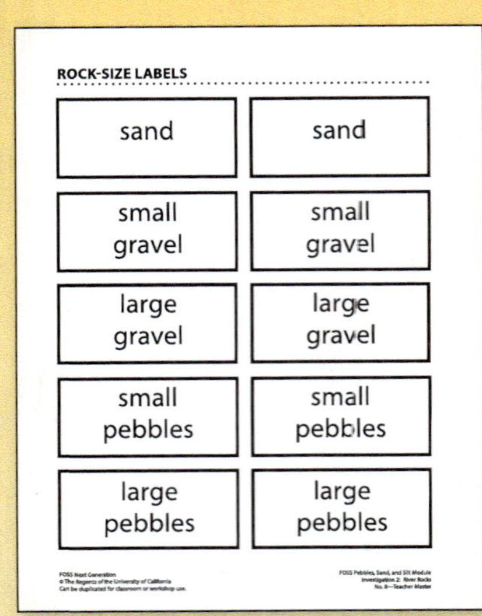

ROCK-SIZE LABELS

sand	sand
small gravel	small gravel
large gravel	large gravel
small pebbles	small pebbles
large pebbles	large pebbles

No. 8—Teacher Master

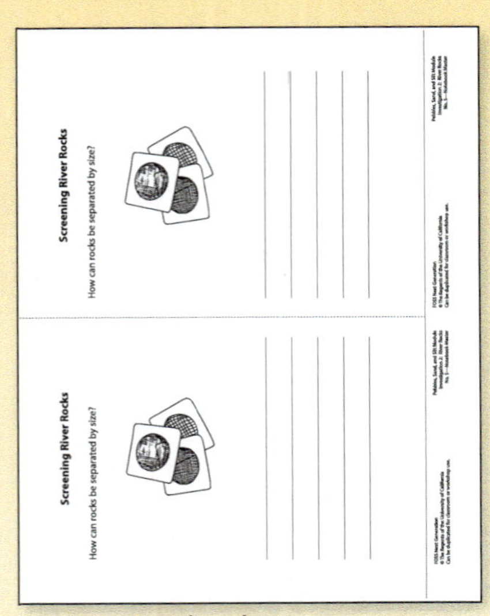

Screening River Rocks

How can rocks be separated by size?

Screening River Rocks

How can rocks be separated by size?

No. 5—Notebook Master

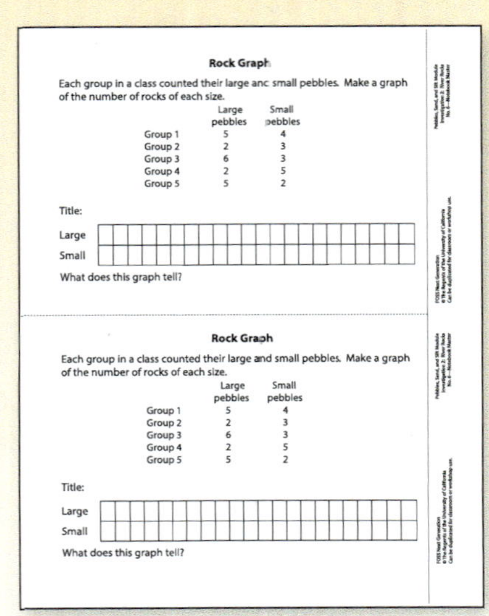

Rock Graph

Each group in a class counted their large and small pebbles. Make a graph of the number of rocks of each size.

	Large pebbles	Small pebbles
Group 1	5	4
Group 2	2	3
Group 3	6	3
Group 4	2	5
Group 5	5	2

Title:

Large

Small

What does this graph tell?

Rock Graph

Each group in a class counted their large and small pebbles. Make a graph of the number of rocks of each size.

	Large pebbles	Small pebbles
Group 1	5	4
Group 2	2	3
Group 3	6	3
Group 4	2	5
Group 5	5	2

Title:

Large

Small

What does this graph tell?

No. 6—Notebook Master

MATERIALS *for*

Part 1: *Screening River Rocks*

For each pair of students

- 1 Set of three screens (small-, medium-, and large-mesh)
- 5 Containers, 1/4 L
- 1 Plastic cup
- 2 Paper plates
- 2 Hand lenses
- 1 Set of *Rock-Size Labels* (See Step 3 of Getting Ready.)
- ❑ 2 Notebook sheet 5, *Screening River Rocks*
- ❑ 2 Notebook sheet 6, *Rock Graph* (optional)

For the class

- 1 Rock mixture (See Step 4 of Getting Ready.)
  - Large pebbles
  - Small pebbles
  - Gravel
  - Sand, unwashed with silt
- 2 Basins
- 1 Metal spoon
- 1 Whisk broom and dustpan
- ❑ 1 Teacher master 8, *Rock-Size Labels*

For embedded assessment

- ❑ • *Performance Assessment Checklist*

❑ Use the duplication master to make copies.

GETTING READY *for*

Part 1: *Screening River Rocks*

1. Schedule the investigation

This part will take one or two active investigation sessions. Students may benefit from additional sessions with the screens. A possible breakpoint for this purpose is suggested after Step 10. Making a graph is an optional math activity and will take another session.

2. Preview Part 1

Students separate a river-rock mixture, using a set of three screens. They discover five sizes of materials: large pebbles, small pebbles, large gravel, small gravel, and sand. The focus question is **How can rocks be separated by size?**

3. Prepare rock-size labels

Each pair of students will need one set of five rock-size labels. Make a copy of teacher master 8 for each group, cut the labels apart into two sets, and keep each set together.

4. Prepare rock mixture

If you are the first person to use the kit, prepare the class rock mixture in a basin. Measure with a 1/4 L container.

2 containers of sand, unwashed with silt

2 containers of gravel

2 containers of small pebbles

2 containers of large pebbles

If the kit has been used before, look for a large zip bag labeled "River Rocks." Each pair of students will need a plastic cup half full of the rock mixture. Spoon the mixture into cups ahead of time.

5. Separate the large-mesh screens

Students use only the large-mesh screen at first, so separate it from each set for distribution in Step 7 of Guiding the Investigation.

6. Plan assessment: performance assessment

In this part, students sort the rock mixtures into size groups using screens with various sized holes. This will provide an opportunity for you to observe students' scientific practices and see how systematic they are in their sorting procedures. Carry the *Performance Assessment Checklist* with you as you visit the groups while they work.

NOTE
Enough materials are provided to conduct this activity with the whole class working in pairs. However, you may want to work with fewer students at a learning center.

NOTE
To prepare for this investigation, view the teacher preparation video on FOSSweb.

NOTE
The washed, clean sand is only used for making sand sculptures in Investigation 3. Use the unwashed sand with silt for making the river-rock mixture.

FOCUS QUESTION

How can rocks be separated by size?

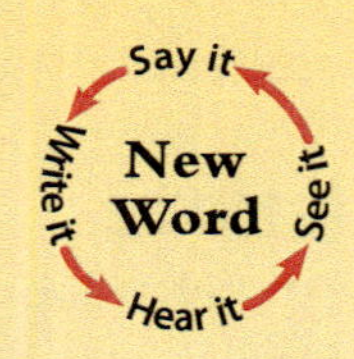

Materials for Step 2

- *Cups of rock mixture*
- *Paper plates*
- *Hand lenses*

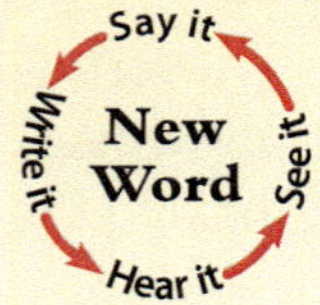

EL NOTE

Make sure students understand that separate means to take apart.

GUIDING *the Investigation*

Part 1: *Screening River Rocks*

1. Introduce the rock mixture

Call students to the rug. Display the image of the river on page 7 of *FOSS Science Resources*. Give students time to discuss with a partner what questions they have about rocks in rivers. Then, show them the rock mixture in the basin. Tell them,

Here are some earth materials that came from the edge of a river, like the one in the photograph. They are called river rocks. This is a ***mixture*** *of river rocks. We will continue our job as geologists to find out all we can about the rocks in this mixture.*

2. Distribute rock mixture

Distribute the materials—one plastic cup half-full of rock mixture, two paper plates, and two hand lenses for each pair.

3. Observe the river rocks

Have students divide the mixture between the two paper plates and observe. Visit students to discuss their observations. Allow about 5 minutes for this observation.

4. Return the mixture to the cup

Ask students to return their rock mixtures to their cups. Demonstrate how to fold a paper plate in half to funnel the rock mixture into the cup.

5. Focus question: How can rocks be separated by size?

When the mixtures are back in the cups, ask a few students for their observations. When students report there are different sizes of rocks, tell them,

When people build highways, sidewalks, and houses, they use rocks like these to construct them. But they need to ***separate*** *the mixture by size. How could we separate our rock mixture?*

Ask the focus question and project or write it on the board.

➤ *How can rocks be separated by size?*

TEACHING NOTE

It is OK to give students all three screens at once and let them use the screens without direction.

6. Introduce the first screen

After students have offered their suggestions, show them the large-mesh screen. Ask,

➤ *How could we use this* ***screen*** *to separate the rock mixture?*

Show them how to put a paper plate under a 1/4 L container and put a screen on top of the container. Tell students that they will have to take turns using the screen.

7. Use the large-mesh screen

Distribute one large-mesh screen and two containers to each pair. Allow a few minutes for screening.

Materials for Step 7

- *Large-mesh screens*
- *Containers*

8. Discuss results

Call for attention or, if necessary, return to the rug. Ask students how they used the screens. If needed ask,

➤ *What happened to the rock mixture that was too large to go through the screen?* [It stayed on top of the screen.]

➤ *Where do the rocks go that are smaller than the screen holes?* [They fall into the container.]

9. Introduce medium- and small-mesh screens

Show students the medium- and small-mesh screens. Compare them to the large screen. Say,

Here are two new screens. This one has medium holes and this one has little holes.

➤ *Can you use these screens to separate the rock mixture into more sizes?*

10. Distribute materials

Distribute the medium- and small-mesh screens and three more containers to each pair. Let students work unguided. They may be very unsystematic in their efforts to separate the mixture. Don't hurry them. Make sure each student gets several chances to screen the mixture.

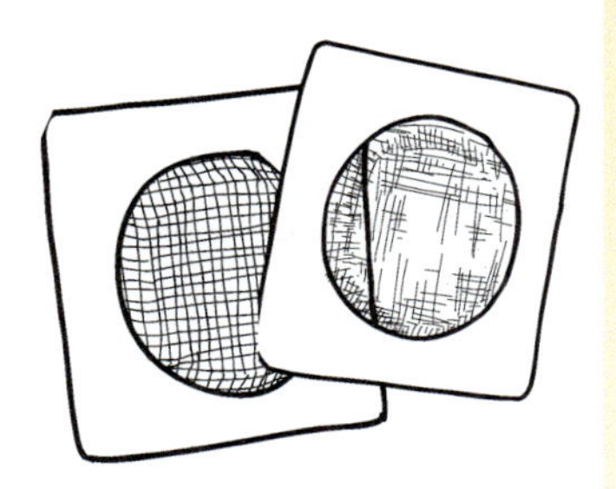

Materials for Step 10

- *Medium- and small-mesh screens*
- *Containers*

POSSIBLE BREAKPOINT

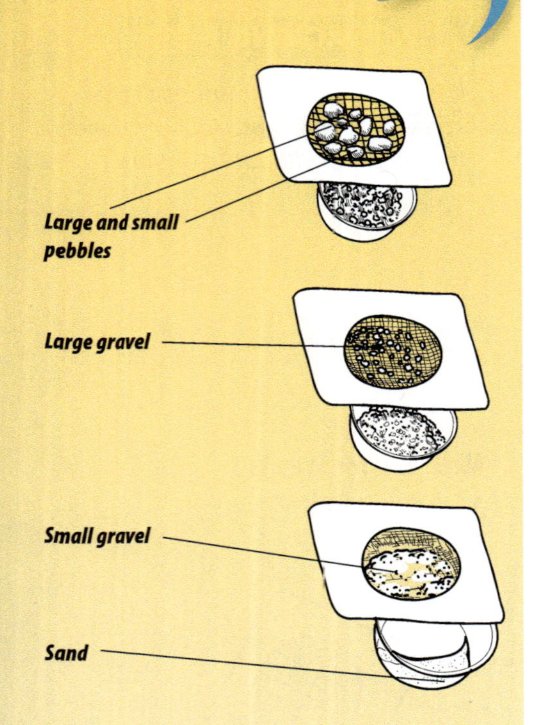

11. Demonstrate three screens

Call students to the rug. Together, plan and then demonstrate the screening procedure.

Note: If the students have experienced the **FOSS Solids and Liquids Module**, they will be familiar with the screens so they should be able to describe how to use them systematically.

a. *Place the large-mesh screen on a container on top of a paper plate. Sift the rock mixture through the screen. Pour the material on top of the screen into an empty container.*

b. *Place the medium-mesh screen on another container and repeat the sifting process with the material that passed through the large-mesh screen.*

c. *Follow the same procedure for the small-mesh screen.*

To guide the discussion, ask,

➤ *How many sizes of rock do we have now?* [Four.]

➤ *Which screens did the smallest pieces go through?* [All three screens.]

➤ *Which screens did the largest rocks go through?* [None.]

12. Use all three screens again

Challenge students to use the three screens to separate their rock mixture into four containers. This separation will result in large and small pebbles in one container, large gravel in another, small gravel in a third, and sand in the last container. Allow 10 minutes.

13. Assess progress: performance assessment

Observe while the groups work and note how systematically they use the screens to conduct the investigation. Conduct 30-second interviews to allow students to explain their methods. Make notes on the *Performance Assessment Checklist*.

What to Look For

- *Students begin with the largest screen, then the medium, and finally the smallest. (Planning and carrying out investigations; scale, proportion, and quantity.)*
- *Students can explain a systematic and logical approach for separating the different sizes, e.g., why they are using the screen with the largest holes first. (Constructing explanations; PS1.A: Structure and properties of matter; patterns.)*

SCIENCE AND ENGINEERING PRACTICES

Planning and carrying out investigations

Constructing explanations

DISCIPLINARY CORE IDEAS

PS1.A: Structure and properties of matter

CROSSCUTTING CONCEPTS

Patterns

Scale, proportion, and quantity

14. Separate large and small pebbles

As students work, go from team to team. As you see a team successfully separate the mixture, ask the team,

➤ *Were the screens useful in separating the mixture?*

➤ *Were there any rocks that did not pass through any of the screens?*

➤ *Can you separate those large rocks into two groups by hand?*

Encourage students to separate the largest rocks into two groups, using the fifth container for the largest size. Ask them to put the rock groups (containers) in order by size.

TEACHING NOTE

This hand separation of pebbles into large and small results in five sizes of rocks in the mixture.

15. Have a sense-making discussion

Call students to the rug. For demonstration, borrow a set of containers from one pair.

Tell students that geologists have names for the different rock sizes. As you identify the sizes, place the appropriate label in the container. Tell students,

The largest rocks you separated are ***pebbles****. There are two sizes of pebbles in this mixture, large pebbles and small pebbles. Ask,*

➤ *How would you describe the size of this rock* [hold up gravel] *compared to pebbles?*

Gravel *is smaller than pebbles. You have two sizes of gravel, small and large. The smallest size is* ***sand****.*

Add these names to the word wall, listing them in order of size: large pebbles, small pebbles, large gravel, small gravel, sand.

Ask,

➤ *How might a large pebble become pieces of gravel in a river?*

If students need scaffolding to answer that question ask,

➤ *What was the effect on the rocks when we rubbed them together?*

➤ *How would this happen in a river?*

Materials for Steps 15–16

- *Sets of labels*

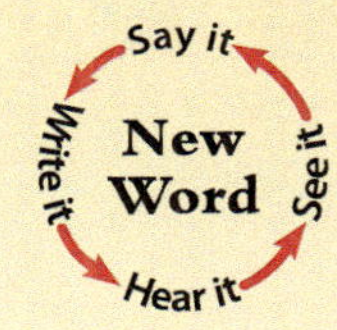

CROSSCUTTING CONCEPTS

Cause and effect

16. Label rock sizes

Distribute a set of five labels to each pair of students. Ask them to return to their tables and put the labels in the container with the rocks of that size. Move from team to team, reviewing their work.

17. Clean up

Have students dump the rocks from their containers into one or two basins. They should brush stray grit on their tables onto the paper plates and then into the basins. Have students bring the other materials to the materials station.

Materials for Step 17

- *Basins*
- *Whisk broom and dustpan*

▶ NOTE

The whisk broom and dustpan can be pressed into service to clean up larger spills.

gravel
mixture
pebble
sand
screen
separate

18. Review vocabulary

Review key vocabulary added to the word wall in this part. One way to do this is to use cloze review. You say a sentence, leaving the last word off, and ask students to answer chorally. Here's an example of cloze review for this part.

➤ *We poured earth materials through different sizes of _____.*

S: Screens.

➤ *The screens helped separate the rocks by _____.*

S: Size.

➤ *The largest rocks we separated are called _____.*

S: Pebbles.

➤ *The smallest rocks we separated are called _____.*

S: Sand.

➤ *The earth material that is bigger than sand but smaller than pebbles is _____.*

S: Gravel.

TEACHING NOTE

Students should describe that sand can go through the screens with the smallest holes. The small gravel can go through screens with medium-sized holes. The largest pebbles stay on top of all the screens.

19. Answer the focus question

Distribute copies of notebook sheet 5, *Screening River Rocks*. Have students glue it into their notebooks and date the page.

Together read the focus question.

➤ *How can rocks be separated by size?*

Then let students work on the entry independently.

20. Create a graph (optional)

Give each student a copy of notebook sheet 6, *Rock Graph*. Read the problem with the class. Be sure students understand that they will use the data to make a graph. Finally, they will determine what information the graph communicates.

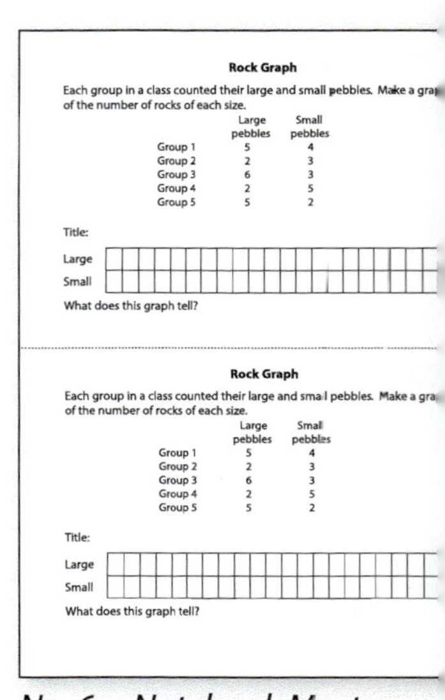

Rock Graph

Each group in a class counted their large and small pebbles. Make a gra
of the number of rocks of each size.

	Large pebbles	Small pebbles
Group 1	5	4
Group 2	2	3
Group 3	6	3
Group 4	2	5
Group 5	5	2

Title:

Large

Small

What does this graph tell?

Rock Graph

Each group in a class counted their large and small pebbles. Make a gra
of the number of rocks of each size.

	Large pebbles	Small pebbles
Group 1	5	4
Group 2	2	3
Group 3	6	3
Group 4	2	5
Group 5	5	2

Title:

Large

Small

What does this graph tell?

No. 6—Notebook Master

SCIENCE AND ENGINEERING PRACTICES

Using mathematics and computational thinking

Notes on the problem. Students have been making concrete and representational graphs with rocks. This problem asks students to apply their understanding of graphs by creating an abstract graph with data that they have not generated. The importance of labeling the graph is embedded, and the activity reinforces the purpose of a graph—to communicate information.

WRAP-UP/WARM-UP

21. Share notebook entries

Conclude Part 1 or start Part 2 by having students share notebook entries. Ask students to open their science notebooks to the most recent entry. Read the focus question together.

➤ *How can rocks be separated by size?*

Ask students to work with a partner to

- share their answers to the focus question.

ELA CONNECTION

This suggested strategy addresses the Common Core State Standards for ELA.

SL 1: Participate in collaborative conversations.

This is a good opportunity to introduce engineering design. Tell students that the screens were designed to solve a problem. Ask them to discuss what the problem was. [How to separate rocks into groups of different sizes.] Next, ask students to discuss other ways this problem can be solved. Explain that to find out which is the best design they would have to try out each one and compare them. Tell students that in the next part, they will use another method to sort rocks by sizes.

MATERIALS *for*

Part 2: *River Rocks by Size*

For each pair of students

- 1 Container, 1/4 L
- 2 *Sand, Gravel, and Pebbles Mat* sheets
- ❑ 2 Notebook sheet 7, *River Rocks by Size*
- 2 Notebook sheet 8, *The Story of Sand* (optional)
- 2 *FOSS Science Resources: Pebbles, Sand, and Silt*
  - "The Story of Sand"

For the class

- 1 Basin with rock mixture (from Part 1)
- 2 Zip bags, 4 L
- 1 Vial, empty
- 1 Vial of sand (from Investigation 1)
- 1 Document camera (optional) ★
- • Transparent tape (optional) ★
- 32 Sheet protectors, clear (optional) ★
- ❑ 1 Teacher master 9, *Sand, Gravel, and Pebbles Mat*
- 1 Big book, *FOSS Science Resources: Pebbles, Sand, and Silt*

For embedded assessment

- ❑ • *Embedded Assessment Notes*

★ Supplied by the teacher. ❑ Use the duplication master to make copies.

No. 7—Notebook Master

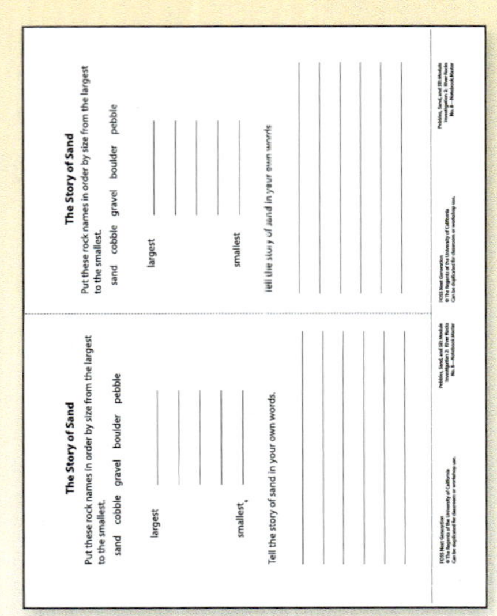
The Story of Sand

Put these rock names in order by size from the largest to the smallest.

sand cobble gravel boulder pebble

largest

smallest

Tell the story of sand in your own words.

No. 8—Notebook Master

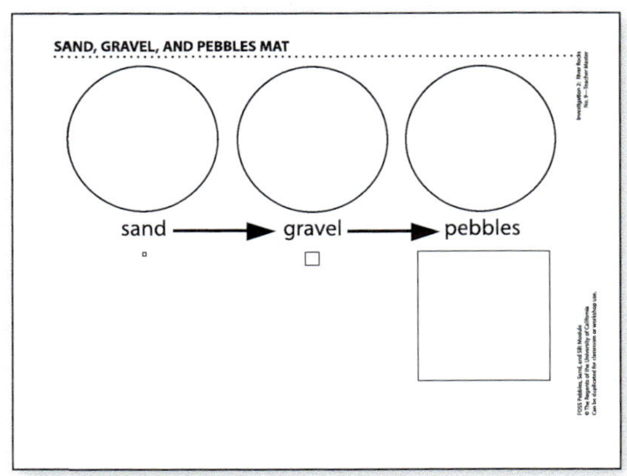

No. 9—Teacher Master

GETTING READY *for*

Part 2: *River Rocks by Size*

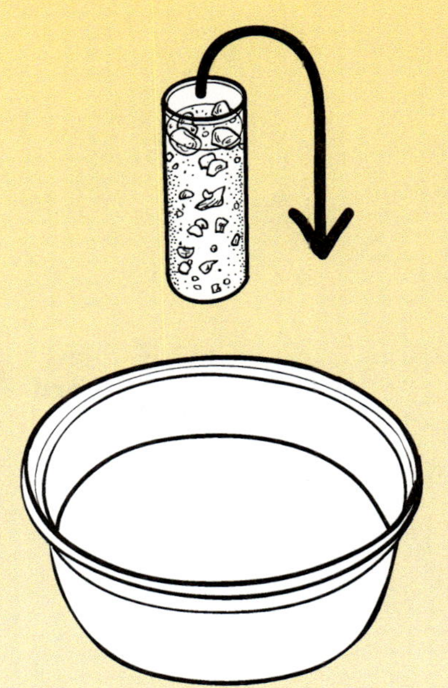

1. **Schedule the investigation**
   This part will take one active investigation and one reading session.

2. **Preview Part 2**
   Students use squares of three sizes as a tool (instead of screens) to seriate rock particles into sand, gravel, and pebbles. The focus question is **How else can rocks be sorted by size?**

3. **Put rock mixture in containers**
   Each pair of students will get a 1/4 L container with a small amount of rock mixture. Measure one vial of the mixture into each container.

4. **Prepare *Sand, Gravel, and Pebbles Mat***
   Each student will need a copy of teacher master 9, *Sand, Gravel, and Pebbles Mat.* The mats can be saved and used by the next class. For protection, mats can be placed in clear plastic sheet protectors.

5. **Plan to tape rock particles in notebooks (optional)**
   Students can tape small pieces of sand and gravel onto their notebook sheets. You might want to circulate to each student and tape the samples down for them.

6. **Plan to use a document camera (optional)**
   A document camera can be used for the demonstration in Step 2 of Guiding the Investigation.

7. **Retrieve the vial of sand from Investigation 1**
   In Investigation 1, Part 1, Step 7, the students contributed sand to a vial. This sand was the result of rubbing rocks together. Retrieve that vial for this part.

8. **Set up a rock-weighing center**
   To give students an introductory experience with a tool to weigh objects, set up a rock-weighing center. Students can find out how many standard or nonstandard units pebbles weigh. See the math extension at the end of this investigation.

9. **Plan to read *Science Resources*: "The Story of Sand"**
   Plan to read "The Story of Sand" during a reading period.

10. **Plan assessment: notebook entry**
    In Step 8, students read "The Story of Sand." Students write a sentence or two to answer the question, "How do rocks change?" Check notebook entries to see that students can explain how rocks can change size from boulders to grains of sand over long periods of time.

GUIDING the Investigation

Part 2: River Rocks by Size

FOCUS QUESTION

How else can rocks be sorted by size?

1. Review rock sizes

Gather students at the rug and review the five sizes of rock particles that students separated from the rock mixture using the screens: large and small pebbles, large and small gravel, and sand. Emphasize that all these particles came from bigger pieces of the same rock.

2. Focus question: How else can rocks be sorted by size?

Ask the focus question and project or write it on the board.

➤ *How else can rocks be sorted by size?*

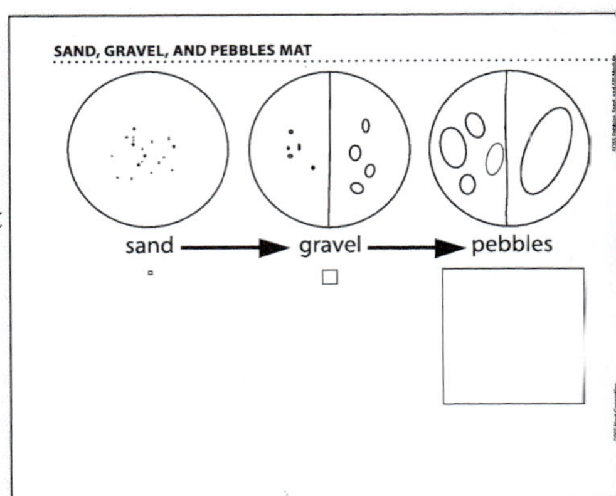

Show students the *Sand, Gravel, and Pebbles Mat*. Tell them that they can use the mat to sort rocks by size. Each pair will get a container with a small amount of rock mixture. Each person gets a mat.

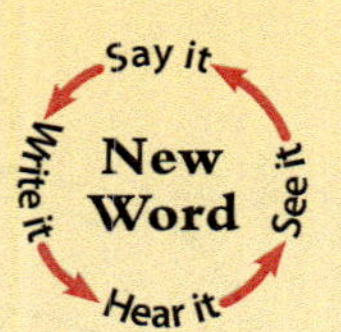

Pick out one rock **particle** from the container. Explain that a particle is a single piece, usually a small piece.

a. *Check to see if the particle will fit inside the square labeled "sand" on the sheet. If the particle fits inside the square, it's sand. Place it in the circle labeled "sand." The square is the same size as the smallest mesh screen we used.*

b. *If the piece doesn't fit in the sand square, try the gravel square. If it fits, place the particle in the "gravel" circle. The square is the same size as the largest mesh screen we used. Both the small and large gravel went through that screen.*

c. *If it doesn't fit in the gravel square, try the pebble square. Place the pebble-sized rocks in the circle labeled "pebble." We didn't have a screen that the pebbles went through.*

3. Distribute materials and begin

Have the Getters pick up one container of rock mixture and two mats for each pair. Move from group to group, challenging students to identify rock sizes that you pull out of their cups.

As they sort, challenge students to divide the pebbles into large and small pebbles and the gravel group into large and small gravel. Students can divide the circle into two halves with a pencil line and separate each group into large and small rocks.

Materials for Step 3

- ***Sand, Gravel, and Pebbles Mat*** *sheets*
- *Containers with rocks*

4. Make a permanent record

Have students use notebook sheet 7 to make a permanent record of the size groups of the rocks. The notebook sheet is a smaller version of the sorting mat that students just used.

Ask students to draw the sand, gravel, and pebbles in the circles to create a representational record. You can have students tape small samples of the actual earth materials of sand and gravel onto the sheet. It is best if students draw the pebbles (they can trace them), as the notebook becomes too hard to close with pebbles inside.

To review the meaning of the graphics on the notebook sheet, ask,

➤ *What do each of the squares show (or mean)? Why are the squares different sizes?* [Each square represents a mesh that the particle will go through. They are like the screens.]

➤ *How are the squares different than the rocks or the screens?* [The squares represent the size of the mesh that a rock will go through.]

Materials for Step 4

- *Tape*

SCIENCE AND ENGINEERING PRACTICES

Developing and using models

Analyzing and interpreting data

5. Clean up

Once all students have identified some of each size of rock, have them return the rocks to the container. Have the Getters return the rock mixture to the materials station.

EL NOTE

For students who need support, provide a sentence frame such as We sorted rocks by _____. First, _____. Next, _____. Then, _____.

6. Store river-rock mixture

At the end of the investigation, store the river-rock mixture in a 4 L zip bag. For added protection, put this bag into a second zip bag. Label the outer bag "River Rocks."

7. Answer the focus question

Have students answer the focus question in their notebooks.

Together read the focus question.

➤ *How else can rocks be sorted by size?*

Then let students work on the entry independently.

TEACHING NOTE

See the ***Home/School Connection*** *for Investigation 2 at the end of the Interdisciplinary Extensions section. This is a good time to send it home with students.*

The Story of Sand

Have you ever looked at one grain of **sand** and thought, "I wonder how it got so small?"

14

Materials for Step 8

- *Vial of sand*

EL NOTE

Explain that a single piece or particle of sand is often called a grain of sand.

ELA CONNECTION

These suggested strategies address the Common Core State Standards for ELA.

RI 8: Describe how reasons support points the author makes in the text.

RF 4: Read with accuracy and fluency to support comprehension.

SL 2: Recount or describe key ideas.

L 4: Determine or clarify the meaning of unknown or multiple-meaning words and phrases.

L 6: Use acquired words and phrases.

READING *in Science Resources*

8. Read "The Story of Sand"

Point out the title of the article and ask students to discuss with a partner what they think the "The Story of Sand" will be about. Ask students what questions they have about sand.

Read the first line of the article aloud to students: "Have you ever looked at one grain of sand and thought, 'I wonder how it got so small?'" Pause and ask students to discuss with a partner how a grain of sand got so small. Ask,

- ➤ *Was the grain of sand always so small?*
- ➤ *Was the grain of sand always at the beach?*

Explain that the "The Story of Sand" will help them answer these questions and they will learn new words to describe rocks of different sizes. Read aloud or have students read the article independently.

Return to page 15 and model how you might figure out what "boulder" means using the context and the photograph. To confirm, tell students they can find the meaning by looking up the word "boulder" in the glossary. Say,

It says here a boulder is a very large rock that is bigger than a cobble. So we know it's a big rock, but what size is a cobble?

Have students look up "cobble" to discover that a cobble is bigger than a pebble. Write the new words on the word wall with an illustration to show the difference in sizes. (See the Background for Teacher section for the size ranges in millimeters.) Tell students that a cobble is about the size of a large potato or apple. Some students may know the term "cobble stone" in cobble stone streets.

Have students pair up and take turns describing how a boulder can end up as a grain of sand using the photographs from pages 15–18. Encourage them to use the rock particle size words that are bolded in the text. Next, have students reread page 18. Explain that this paragraph summarizes the main idea of the article. Have students look for evidence from the previous pages that supports the author's explanation of weathering.

Give students time to look at and discuss the sand samples from different places on page 19. Have them imagine what the rocks looked like when they were bigger.

Finish the reading session by having students read the last page and share their experiences with sand. Ask students why they think the author used an exclamation point at the end of the sentence.

9. Have a sense-making discussion

To check for understanding, rephrase the questions asked earlier and have students discuss them in their groups. Encourage them to use the text to support their answers.

- *What caused a grain of sand to get so small?* [Weathering caused large rocks to break into small rocks.]
- *Was a grain of sand always so small?* [It was once a part of a larger rock.]
- *Was a grain of sand always at the beach?* [It could have moved by wind or water from another place.]
- *When we rubbed rocks together, what was the effect?* [Sand.]
- *What was the effect on the size of the two rocks?* [The surface of the rocks was rubbed off; they got a little smaller.]

This might be a good time to bring out the vial of sand the class made by rubbing the three rocks together in the first investigation.

Use these questions to deepen students' understanding. Have students read the article again independently or with a partner. Have students respond to the second question in their notebooks.

- *Where do rocks that we find on the ground come from? Were they always where we found them? Where could they have come from?*
- *How do rocks change?* [Discuss and list the possibilities and have students respond to this prompt in their notebooks: boulders break off mountains, water moves rocks and bumps them together, waves crash onto big rocks, etc.]
- *How long does it take for a boulder to change into sand?* [It takes a very long time.]

Reread pages 15 to 18 using the interactive eBook and showing the videos on each page.

SCIENCE AND ENGINEERING PRACTICES

Obtaining, evaluating, and communicating information

ELA CONNECTION

These suggested strategies address the Common Core State Standards for ELA.

RI 1: Ask and answer questions to demonstrate understanding.

SL 5: Add drawings or other visual displays to recounts of experiences.

CROSSCUTTING CONCEPTS

Cause and effect

Stability and change

10. Assess progress: notebook entry

Have students answer the discussion question (Step 9) "How do rocks change?" Students can draw a picture, write a few sentences, or both. Collect their notebooks after class and check their answers.

SCIENCE AND ENGINEERING PRACTICES

Constructing explanations

What to Look For

- *Students explain that when rocks are rubbed together, they change because small pieces can fall off.*
- *Students explain that smaller pieces of rock are often transported by wind or water.*

11. Review in science notebook (optional)

Have students respond to the prompts on notebook sheet 8 to answer the focus question. (These prompts appear at the end of the reading.) The rock order by particle size is boulder, cobble, pebble, gravel, and sand.

As an extension, have students pretend they are a particle of sand. Have them write a narrative describing how they got to be so small.

ELA CONNECTION

This suggested strategy addresses the Common Core State Standards for ELA.

W 3: Write narratives.

EL NOTE

Point out the compare and contrast or sequence language structures students should use and encourage them to write in complete sentences. As a scaffold, provide sentence frames such as:

Compare and contrast:
A _____ is smaller than a _____.
A _____ is bigger than a _____.

Sequence:
First, _____.
Next, _____.
Then, _____.
Finally, _____.

WRAP-UP/WARM-UP

12. Share notebook entries

Conclude Part 2 or start Part 3 by having students share notebook entries. Ask students to open their science notebooks to the most recent entry. Read the focus question together.

➤ *How else can rocks be sorted by size?*

Ask students to work with a partner to

- share their answers to the focus question;
- describe what they drew.

Encourage students to explain the weathering process in terms of cause and effect. [Wind and water cause rocks to move; rocks bumping into each other causes them to break apart.]

CROSSCUTTING CONCEPTS

Cause and effect

MATERIALS *for*

Part 3: *Sand and Silt*

For each student

- 1 Vial with cap
- 1 Paper plate
- 1 Plastic spoon
- 1 Hand lens
- 1 Self-stick note
- ❑ 1 Notebook sheet 9, *Sand and Water Drawing*

For the class

- • Sand, unwashed with silt, 1 L (about 1 kg)
- • Dry powdered clay (See Step 5 of Getting Ready.)
- 9 Plastic cups
- 8 Containers, 1/4 L
- 1 Basin
- 1 Bottle brush
- 1 Pitcher
- 8 Vial holders
- • Water ★
- • Paper towels ★
- • Transparent tape ★
- 1 Document camera or other projection system (optional) ★

For embedded assessment

- ❑ • *Embedded Assessment Notes*

★ Supplied by the teacher.

❑ Use the duplication master to make copies.

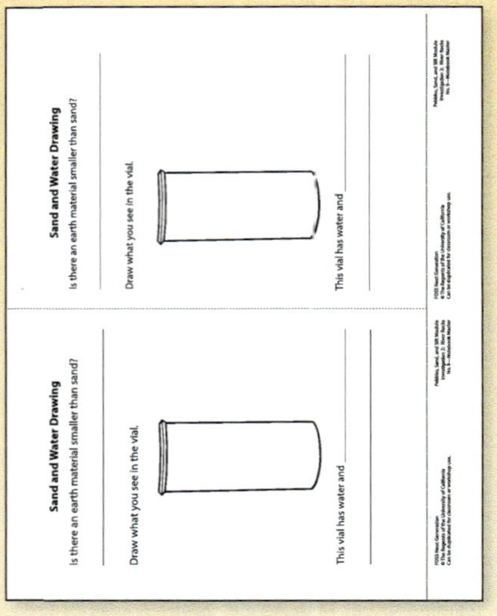

Sand and Water Drawing

Is there an earth material smaller than sand?

Draw what you see in the vial.

This vial has water and ___

Sand and Water Drawing

Is there an earth material smaller than sand?

Draw what you see in the vial.

This vial has water and ___

No. 9—Notebook Master

GETTING READY *for*

Part 3: *Sand and Silt*

1. Schedule the investigation

Schedule two active investigation sessions on consecutive days for this part. Students set up mixtures of sand and water in the first session, and in the second session, record the results once their mixtures have settled overnight.

2. Preview Part 3

Students take a close look at sand and separate sand particles from silt particles, which are smaller than the sand, by mixing the sand with water and allowing the particles to settle. They observe that the sand settles to the bottom and the silt forms a layer on top of the sand. The focus question is **Is there an earth material smaller than sand?**

3. Prepare the vials

Fill a vial two-thirds full with sand for each student. The caps are distributed separately. Have transparent tape ready to secure the self-stick notes to the vials.

Each group of students will place their vial of sand and water in a group vial holder for storage overnight. Plan where you will store the eight vials holders so they remain undisturbed for a day.

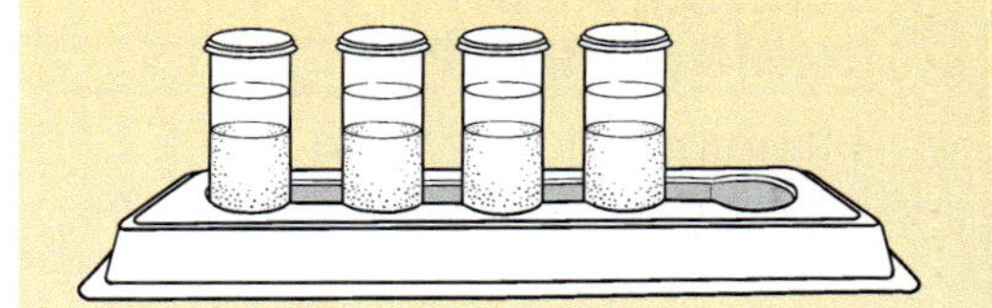

4. Prepare for water

Use a pitcher as a water supply. Pour water into a cup and use the cup to fill students' vials in Step 8 of Guiding the Investigation.

5. Plan to save sand and add silt

At the end of the activity, students will dump their wet sand into a basin. Let the sand dry out to be reused. If an observable layer of silt doesn't appear after the sand dries, mix about 1/4 cup of dry powdered clay into the dry sand before storing it for reuse. Unwashed sand contains silt. Adding dry clay will simulate unwashed sand.

6. Plan for document camera or projection system

If you would like to project the notebook sheet to orient students to its use, arrange to use a document camera or projection system.

7. Prepare for cleanup

You will need warm water and a bottle brush to clean out the vials.

8. Plan assessment: notebook entry

In Step 16 students answer the focus question. Check students' abilities to state a claim and identify the evidence that supports their thinking (engaging in argument from evidence).

FOCUS QUESTION

Is there an earth material smaller than sand?

Materials for Step 2

- *Vials of sand, without caps*

SCIENCE AND ENGINEERING PRACTICES

Planning and carrying out investigations

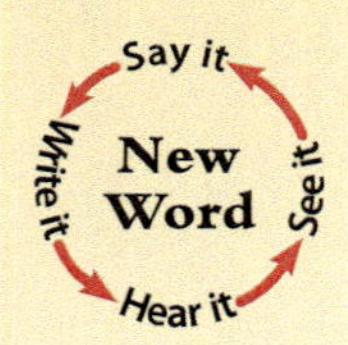

Materials for Steps 4–5

- *Paper plates*
- *Plastic spoons*
- *Hand lenses*

GUIDING *the Investigation*

Part 3: *Sand and Silt*

1. Review sand

Call students to the rug. Show them one of the vials of sand. Ask,

➤ *What do we call rocks of this size?*

Identify it as sand. Remind students that sand was the smallest particle of rock in the river-rock mixture.

Tell students that they will each get a vial of sand to observe. They should look carefully at the sand and describe its properties (appearance and texture).

2. Distribute sand vials

Send students to their tables. Have Getters get a vial of sand for each student in their group. Allow about 5 minutes for free exploration.

3. Introduce plate shaking

Call for attention. Suggest that students use a paper plate and a spoon to find out more about the sand. Demonstrate the procedure.

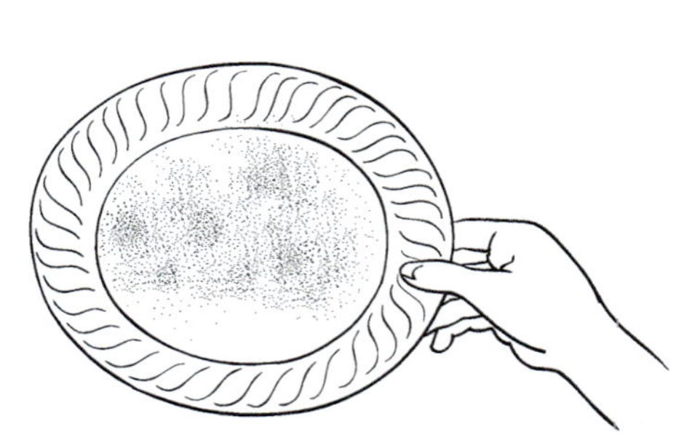

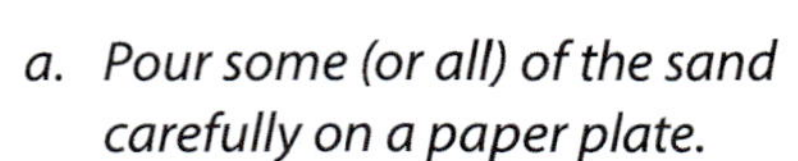

a. *Pour some (or all) of the sand carefully on a paper plate.*

b. *Put the plate on the table.*

c. ***Shake** the plate gently back and forth on the table.*

4. Begin shaking sand on plates

Distribute the plates and spoons. Have students put sand on the plates and begin shaking. Spoons can be used to move the sand around. As you move from group to group, point out how the sand is separating. Allow 5 minutes for shaking.

5. Distribute hand lenses

As students work, distribute a hand lens to each student. Show them how to place just a pinch of sand in the vial, place the hand lens on the top of the vial, and observe grains of sand on the bottom of the vial.

With a lens on top of the vial, sand in the bottom of the vial will be in focus.

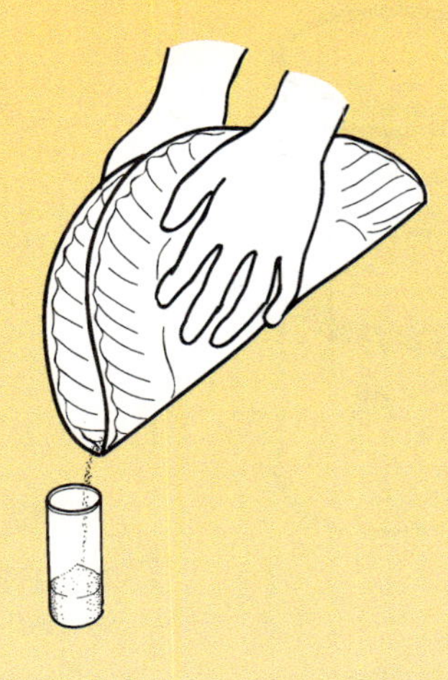

6. Discuss observations of sand

Have students leave the materials at the table and return to the rug. Ask them to describe what they noticed about the sand—how it felt and sounded, different colors, sizes. Ask,

➤ *Are the sand **particles** all the same size?* [No.]

➤ *What do the sand particles look like?* [Little pieces of rock.]

Again, make the connection to the particles of rock students rubbed off the three rocks in the first investigation. Add new words that students use to describe the sand to the word wall.

7. Introduce sand and water in vials

Ask students what might happen if they mix sand and water. Demonstrate this procedure.

a. Use the paper plate as a funnel to put the sand back in the vial.

b. Watch as you fill the vial with water.

c. Without touching the vial, observe what happens to the water and sand.

TEACHING NOTE

Pour the water into the vial quickly so that students can watch the water level in the vial go down and see that the result is an air space of about 1 cm at the top of the vial.

8. Add water to vials

Have students return to their tables. When they have the sand in the vials, fill each vial to the top with water. Be sure the water comes all the way to the top. Visit the groups and ask,

➤ *I filled the vial to the top with water. What happened to the water level?* [It went down.]

➤ *What happened to the water that was poured on the sand?* [It went into the air spaces between the pieces of sand.]

Materials for Steps 8–9

- *Caps for vials*
- *Pitcher of water*
- *Cup*

9. Shake the closed vial

Tell students that each person will get a cap to put on the vial. Holding the vial cap and bottom, they can shake the vial to mix the water and sand. Distribute the caps and allow a few minutes for shaking the vials. Visit the groups and ask,

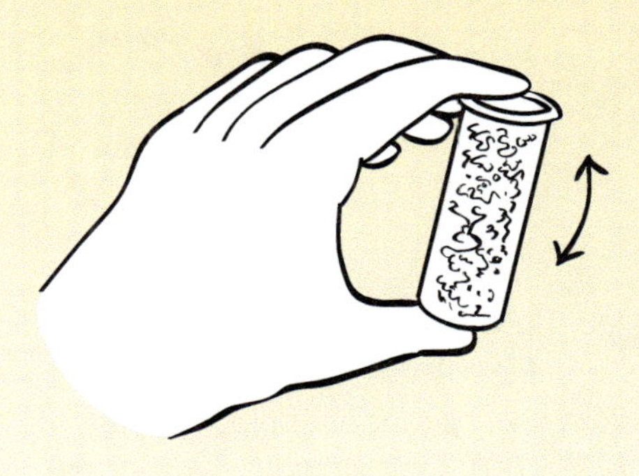

➤ *What happened to the sand and water when you shook the vial?*

➤ *Was the water clear after you shook the vial? Why not?*

10. Let sand and water settle overnight

Distribute self-stick notes to students to label their vials. Distribute a vial holder to each group of students and identify where to place the holders overnight. Make it clear that students should *not* shake the vials the next day. Ask students to predict what might happen in the vials overnight.

Materials for Step 10

- *Vial holders*
- *Self-stick notes*
- *Tape*

B R E A K P O I N T

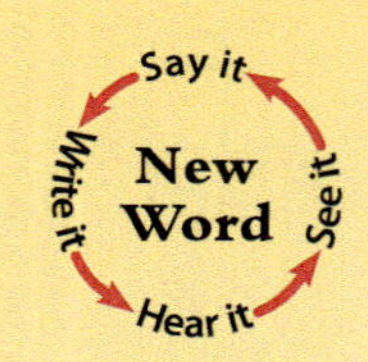

Materials for Steps 12–13

- *Vial holders containing vials with sand and water*
- ***Sand and Water Drawing*** *sheets*

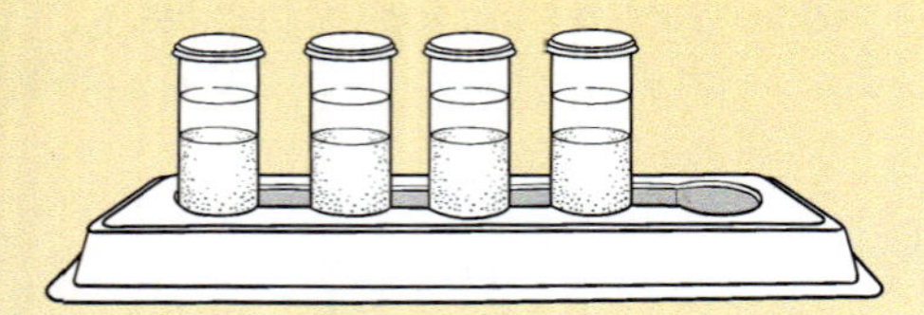

SCIENCE AND ENGINEERING PRACTICES

Analyzing and interpreting data

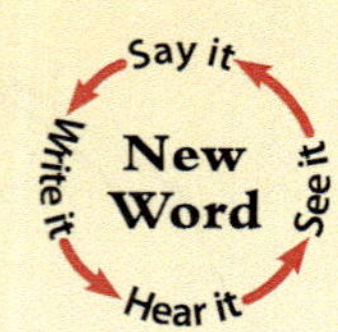

TEACHING NOTE

There will probably be some clay in the top layer as well, but for now it will be referred to as silt.

CROSSCUTTING CONCEPTS

Scale, proportion, and quantity

11. Focus question: Is there an earth material smaller than sand?

Call students to the rug. Ask the focus question and project or write it on the board.

➤ *Is there an earth material smaller than sand?*

Hold up one of the vials of sand and water set up the day before. Show them the notebook sheet to use for drawing the sand vial. Explain that each student should draw what he or she sees in the vial now that it has **settled**.

If you have a document camera or other projection system to display the notebook sheet, this would be the time to use it.

12. Retrieve vial holders

Emphasize that students should be careful not to shake or disturb the sand in the vials. They want to observe the settled vials. Have Getters retrieve the vial holders and bring them to the tables for students to observe. Each student should carefully remove his or her vial from the holder.

13. Draw results of settling

After a couple of minutes of observing, distribute a *Sand and Water Drawing* sheet to each student. Allow 10 minutes for drawing.

14. Have a sense-making discussion

Have students share their observations. Use these questions to guide the discussion.

➤ *What do you see in the vials?*

➤ *How many **layers** do you see?*

➤ *Where's the sand?*

➤ *What do you see on top of the sand?*

➤ *What do you think you would find on the bottom of a pond or lake?*

15. Introduce *silt*

Tell students,

*The sand **sinks** in the water. The sand sank to the bottom of the vial. The layer of material on top of the sand is called **silt**. Silt is a particle of rock much smaller than sand. It's the rock size that mud is made of.*

Point out that, if they shake the vial even a little, the silt goes back into the water.

16. Answer focus question

Tell students to label their drawings to identify the layers they see in the vial. Discuss what the layers should be called: sand, silt, and water. Ask them what they should label the space between the top of the water and the vial cap. [Air.]

Restate the focus question.

➤ *Is there an earth material smaller than sand?*

Have students answer the question by stating a claim, such as Yes, there is an earth material smaller than sand. I know this because ____ .

SCIENCE AND ENGINEERING PRACTICES

Engaging in argument from evidence

17. Assess progress: notebook entry

After class, review students' answers to the focus question.

What to Look For

- *Students answer that there is an earth material smaller than sand (silt).*
- *Students use their drawings of the layered vial as evidence that there are earth materials smaller than sand (the top layer of silt).*

18. Feel the texture of silt

Give each group a cup. Instruct them to carefully pour off the water in the vials into a cup. The sand and silt should remain in the vial. Tell them to carefully touch the top layer of silt. If they can, have them take some silt out of the vial and rub it between their fingers. Add any words they use to describe the silt to the word wall.

Materials for Step 18

- *Cups*

19. Review vocabulary

Review key vocabulary added to the word wall in this part dealing with the behavior of small earth particles. Include the word *particle* in the review. One way to do this is to use cloze review. You say a sentence, leaving the last word off, and ask students to answer chorally. Here's an example of cloze review for this part.

layer
particle
settle
shake
silt
sink

➤ *Each tiny piece of sand is called a ____.*

S: Particle.

➤ *We did this to the vial to mix the sand and water. ____*

S: Shake.

➤ *When the vials settled overnight, we could see the ____.*

S: Layers.

➤ *The layer on top of the sand is called ____.*

S: Silt.

20. Clean up

Have students bring their vials to the cleanup area. Add a little water to the vials, shake, and dump the entire vial contents into a basin. The sand will be collected, dried, and saved for use again. The water should be collected and recycled by watering a plant indoors or outdoors; do not pour the water down the sink drain.

A bottle brush is available for further cleanup. Place the clean vials and cups upside down on some paper towels to dry.

Students should wash their hands at the end of this activity.

Let the sand dry out to be reused. If an observable layer of silt doesn't appear after the sand dries, mix about 1/4 cup of dry powdered clay into the dry sand before storing it for reuse. Unwashed sand contains silt. Adding dry powdered clay will simulate unwashed sand.

WRAP-UP/WARM-UP

21. Share notebook entries

Conclude Part 3 or start Part 4 by having students share notebook entries. Ask students to open their science notebooks to the most recent entry. Read the focus question together.

➤ *Is there an earth material smaller than sand?*

Ask students to work with a partner to

- share their answers to the focus question;
- describe their labeled drawing.

Encourage students to review the process they used to separate the silt from the sand. Ask if they notice any patterns in how the earth materials separate [smaller particles are on the top]. They can also make connections to how sand and water interact in still water such as lakes and ponds.

ELA CONNECTION

This suggested strategy addresses the Common Core State Standards for ELA.

SL 4: Recount an experience.

MATERIALS *for*

Part 4: *Exploring Clay and Landforms*

For each student

- 1 Cube of potter's clay, 2–3 cm (See Step 3 of Getting Ready.)
- 1 Vial with cap
- 1 Self-stick note
- ❏ 1 Notebook sheet 10, *Clay and Water Drawing*
- ❏ 1 Notebook sheet 11, *Rocks in Bottle Drawing*
- 1 *FOSS Science Resources: Pebbles, Sand, and Silt*
  - "Rocks Move"
  - "Landforms"

For each group

- 2 Plastic cups
- 1 Self-stick note

For the class

- 1 Vial
- 1 Cube of clay, 2–3 cm
- 1 Basin
- 1 Bottle brush
- 1 Pitcher
- 8 Vial holders
- • Paper towels ★
- • Water ★
- 1 Large knife or piece of fishing line ★
- • Transparent tape ★
- • Paper plates (for embedded assessment)
- • Sand and clay (for embedded assessment)
- 1 Big book, *FOSS Science Resources: Pebbles, Sand, and Silt*
- 1 Computer with Internet access ★

For embedded assessment

- ❏ • *Embedded Assessment Notes*

For benchmark assessment

- ❏ • *Investigation 2 I-Check*
- • *Assessment Record*

★ Supplied by the teacher. ❏ Use the duplication master to make copies.

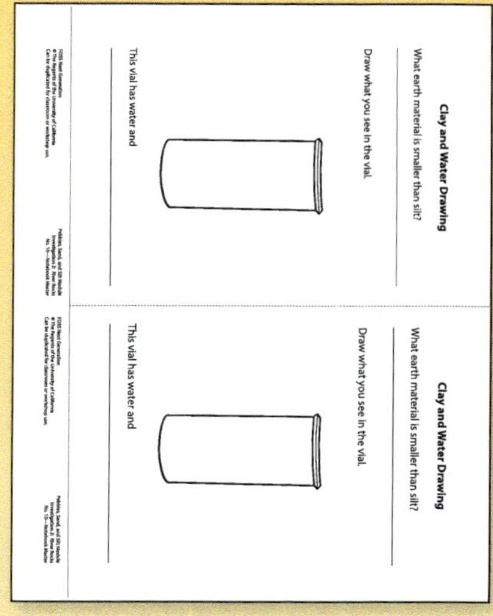

No. 10—Notebook Master

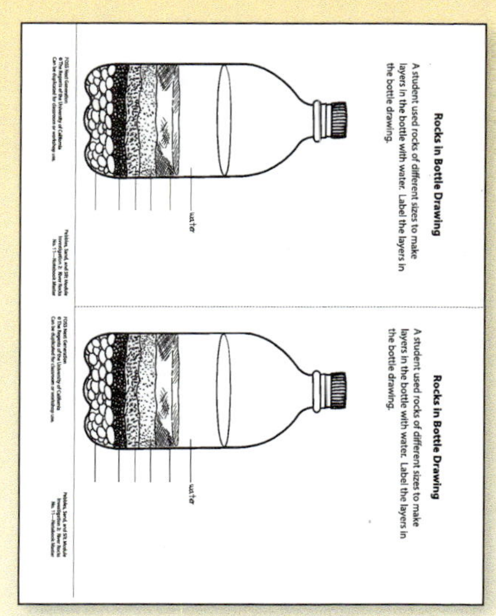

No. 11—Notebook Master

GETTING READY *for*

Part 4: *Exploring Clay and Landforms*

1. Schedule the investigation

This part will take five sessions— two short active investigation sessions on different days for working with clay, two sessions for readings and the video, and one for the I-Check.

2. Preview Part 4

Students investigate the properties of the smallest rock particles, clay. They read about and view a video about ways that wind and water move and shape the land. Students compare the time it takes to change the surface of the land. The focus questions are **What earth material is smaller than silt?** and **how does water and wind change landforms?**

3. Cut clay into cubes

Each student will get a cube of potter's clay about 2–3 cm (1") square. Use a large knife or fishing line to cut the potter's clay into slabs 2–3 cm thick. Cut each slab into 2–3 cm cubes. Make one cup for each group, and put enough pieces of clay into the cups so each student gets one cube of clay.

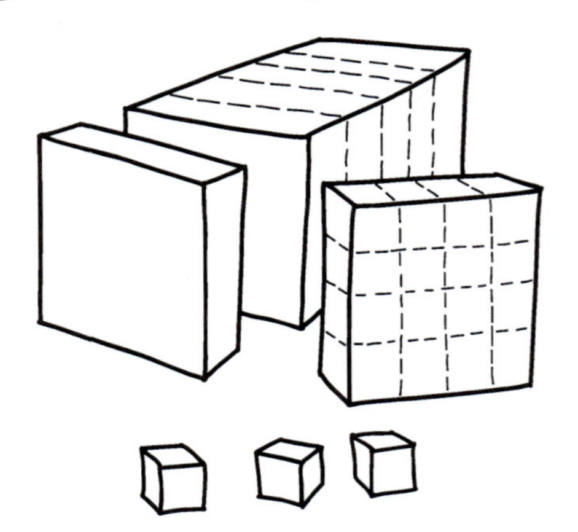

4. Plan for clean up

We suggest that students mold clay on the tabletops and then clean the desks with a damp sponge. If you have plastic surfaces for working clay, use them here. At the end of the activity, students dump their clay and water into a basin. After it settles, pour off the water, and allow the clay to dry somewhat. Recycle it into the clay supply for this module.

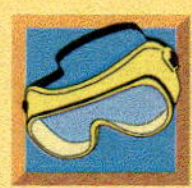

5. **Plan for safety**
Be aware of any allergies that students in your class might have. Students will be working with clay that can result in dust particles. Caution students not to rub their eyes with clay on their hands.

6. **Plan to read *Science Resources*: "Rocks Move"**
Plan to read and discuss "Rocks Move" during a reading period after completing this part. This reading will activate prior knowledge about how wind and water moves rocks and prepare the students for the video and readings to come.

7. **Preview the video**
Preview the video, *All about Land Formations* (duration 21 minutes.) The video reviews some landforms vocabulary already discussed in this investigation and introduces new words to describe landforms formed by weathering (the breaking down of rock) and erosion (the movement of rock and soil to new locations to create new landforms). The main idea from the video is that all of Earth's surfaces are constantly changing, but those changes are small and go unnoticed for thousands of years. This is in contrast to the rapid changes to Earth's surface from the eruptions of volcanoes or the powerful force of water during strong storms.

The video will provide a number of opportunities for students to compare models to what actually happens.

The videos includes these chapters:

- Introduction
- Mountains, mesas, buttes
- Water and land
- Investigation: Making hills and valleys
- Weathering and deposition
- Conclusion

The link to this video for teachers is in the Resources by Investigation on FOSSweb.

8. **Plan to read *Science Resources*: "Landforms"**
Plan to view and discuss "Landforms" during a reading period after viewing the video. The photos will review some of the landforms introduced in the video—mountain, plateau, mesa, butte, valley, canyon, delta, plain, and sand dune.

9. Plan assessment: notebook entry

In Step 27, students make models of sand on paper plates. Students draw landform models in their notebooks. Check that students can draw models that look like the landform they labelled.

You will need to provide paper plates, sand, and clay for students to make models of the landforms as part of this embedded assessment. Plan ahead for how you will conduct this assessment.

10. Plan assessment: I-Check

Plan to give *Investigation 2 I-Check* at the end of the investigation. Read the items aloud to the whole class, and have students answer independently. Review students' responses using the What to Look For information in the Assessment chapter. Use assessment master 3, *Assessment Record,* to record students' responses.

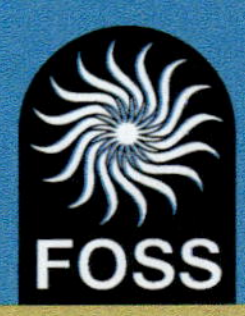

GUIDING *the Investigation*

Part 4: *Exploring Clay and Landforms*

1. Explore the clay

Call students to the rug. Review by asking,

➤ *Which is the smallest size of rock we've worked with so far?* [Silt.]

2. Focus question: What earth material is smaller than silt?

Ask the focus question and project or write it on the board.

➤ *What earth material is smaller than silt?*

Tell students that you have a new earth material for them to observe today that might answer this question. Explain that each student will get a cube of the material (don't call it clay yet), and they should find out all they can about it.

Have students move to their tables. Distribute the cups of clay cubes to each group. Let students explore the clay for about 5 minutes. Caution them not to rub their eyes while working with this earth material.

3. Discuss observations

Have students put down the material. Ask them to share their observations. Guide the discussion with these questions.

➤ *How does it feel?*

➤ *What can you do with it?* [Shape it.]

➤ *What's happening to your hands as you work with this material?* [They are covered with powder or dust.]

4. Introduce the smallest particle-size—clay

Identify that the earth material is clay. Tell students,

***Clay** is made of pieces of rock, even smaller than silt. These pieces are really, really small. The dust on your hands is actually dry clay particles.*

5. Divide clay cubes

Demonstrate as you explain how to divide a clay cube into balls.

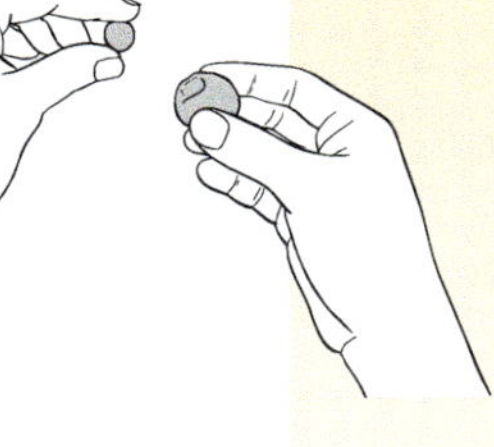

a. *Pinch off a small piece of clay from a cube. The small piece should be about the size of a large pea (less than 1 cm or 1/2").*

b. *Roll both pieces into balls.*

c. *The larger balls of clay will sit in the open cup overnight to find out what happens to them. Each group will share a cup.*

FOCUS QUESTION

What earth material is smaller than silt?
How does water and wind change landforms?

Materials for Step 2

- *Clay cubes in cups*

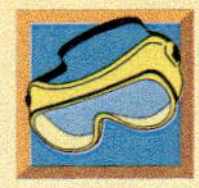

TEACHING NOTE

Tell students that this is potter's clay, which is different from modeling clay or dough clay.

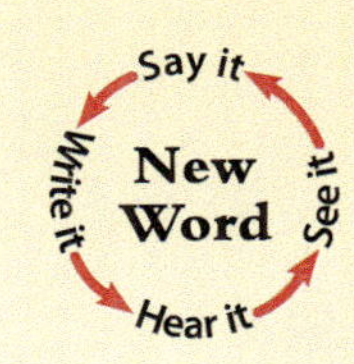

SCIENCE AND ENGINEERING PRACTICES

Planning and carrying out investigations

Materials for Step 7

- *Vials without caps*
- *Cups of water*

Materials for Step 8

- *Vial caps*

TEACHING NOTE

It's OK to shake the student vials often during the first day, but no shaking on the second day. The teacher vial should not be shaken.

Materials for Steps 9–10

- *Self-stick notes*
- *Paper towels*
- *Tape*
- *Vial holders*

6. Demonstrate procedure

Ask students what they think will happen if they put the pea-sized ball of clay in a vial with water. Demonstrate as you explain.

a. *Put the small ball of clay in a vial.*

b. *Carefully pour water from a cup into the vial. Fill the vial almost to the top but do not shake the vial.*

c. *Watch what happens to the clay in the water for a minute or two.*

7. Put clay in vials and add water

Distribute one vial (without cap) to each student. Have students follow the procedure you demonstrated. Distribute a cup of water to each group to share. As students observe the clay ball in the vial or water, ask them to describe what they see.

➤ *What's happening to the clay?* [Stays to the bottom of the water.]

➤ *Is the water changing?* [Not really.]

8. Shake the vials

Supply the caps for the vials when students have completed their first observation. Challenge them to shake the vial until the clay balls disappear. (They probably won't be able to get the clay to disappear.) Ask them to determine whether the clay ball is getting smaller and to describe changes in the water.

9. Store vials overnight

Tell students that they will let the vials sit overnight. Point out that you didn't shake your vial; you'll let your vial sit overnight without shaking so they can compare it with their vials tomorrow.

Distribute a self-stick note to each student to label his or her vial. Provide each group with one additional self-stick note to label the plastic cup containing the group's large clay balls.

Have students put their vials in a vial holder and have the Getters bring the holders and the plastic cups to the storage location or leave them on their desks overnight.

10. Clean up

Have the Getters get damp paper towels for cleaning the clay from the tables. Students should wash their hands with warm water.

BREAKPOINT

11. Review the work with clay
Call students to the rug. Review the work with clay—in water and in air.

12. Observe the dry clay
Distribute the plastic cups with the clay balls. Let students observe and squash the clay balls for a few minutes. The clay may be as hard as a rock or crumbly.

TEACHING NOTE

The dryness of the clay balls depends on the temperature and humidity.

13. Add water
Ask students how they could get the clay soft and pliable again. If they don't suggest it, propose adding water. Have all students in a group put their dry clay balls in the plastic cup. Cover the clay with a little water and put them to one side as students continue with the activity.

Materials for Steps 12–13

- *Clay balls in cups*
- *Pitcher of water*

14. Observe the clay vials
Distribute the vials in the holders. Remind students not to shake them. Have them observe the vials and describe what they see. [The water may still be milky-looking because of suspended clay. There will be a layer of clay on the bottom of the vial.]

Ask students to compare their vials with the vial with water and clay that you left overnight without shaking. Have them also compare what happened in the sand vial after it sat overnight. [The silt settled on top of the sand; the water was much clearer.]

Materials for Step 14

- *Vials of clay*

15. Answer the focus question
Distribute a copy of notebook sheet 10, *Clay and Water Drawing*, to each student. Ask students to draw the clay vials and label the layers and answer the focus question. Have students glue the sheet into their notebooks.

16. Discuss drawings
When students have completed their drawings, ask,

➤ *What was the same in the sand and clay vials?*

➤ *What was different in the two vials?*

SCIENCE AND ENGINEERING PRACTICES

Analyzing and interpreting data

17. Observe clay in water
Direct students' attention to the clay balls in the plastic cups. Ask them to describe what they see. By now the clay balls will probably have crumbled into a mushy layer on the bottom of the cup.

TEACHING NOTE

The view from the top of the cup is especially interesting.

Materials for Step 18

- *Basin*
- *Paper towels*
- *Bottle brush*

18. Recycle the clay

Have students shake up their vials well and dump the contents of their vials and the plastic cups into a basin. Tell them that you will let the basin sit overnight. Have them rinse the vials and caps and turn them upside down on paper towels. It is OK if some clay is left in the vials. It will dry up and release from the vials later.

Place the basin with the water and clay where it can be observed as the water evaporates. When most of the water has evaporated, you can repackage the clay for another use.

clay
silt
sand
gravel
pebble
cobble
boulder

19. Review vocabulary of earth materials

Working with the class, make a list of the earth materials from smallest to largest particle size.

20. Distribute *Rocks in Bottle Drawing*

Distribute a copy of notebook sheet 11, *Rocks in Bottle Drawing*, to each student. Tell students,

A student used rocks of different sizes to make layers in a bottle with water. Label the layers in the bottle drawing.

Ask students to label the layers and to write or dictate a sentence about the picture.

Look to see if students use proper vocabulary in labelling their drawings. Also check to see if they label the layers with the largest particle (pebbles) on the bottom and the smallest (clay) on the top. Some students may break down the gravel into large and small and the pebbles into large and small and have sand on the top layer with no silt or clay in the bottle. That's OK as long as the seriation from large to small is correct.

SCIENCE AND ENGINEERING PRACTICES

Developing and using models
Constructing explanations

READING *in Science Resources*

21. Read "Rocks Move"

Ask students if they think rocks stay in the same place. Have them share their ideas about how rocks move with a partner. Remind students of the article they read about sand and review how rocks moved in that reading.

Introduce "Rocks Move" to students, explaining that, like the story of sand, each of the pictures in the article has its own story.

Ask for a volunteer to read the text aloud.

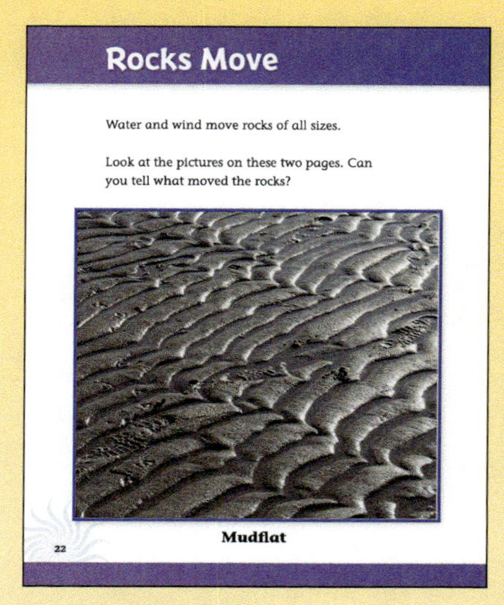
Rocks Move

Water and wind move rocks of all sizes.

Look at the pictures on these two pages. Can you tell what moved the rocks?

Mudflat

22

22. Discuss the reading

Have students share their ideas of what moved the rocks in each of the pictures with a partner. Draw a table on the board or chart paper to help students compare the movement of rocks in the three pictures. Call on volunteers to share what they discussed with their partners. These should include the following key points.

- Water and wind seem to have the same effect on the sandy beach and the mudflat.
- A current in a body of water that once covered the mudflat but has now receded caused the ripples in the mudflat.
- Look at the plants on the beach to identify the direction of the wind.
- In the washout, water rushed down the hillside, carrying sand and silt to the road. Without the smaller particles, the larger rocks couldn't hold together, and the hillside slid.

	What moved the rocks?	What size are the rocks?	What were the results?
Washout	a lot of water, probably a storm	soil—contains clay, silt, sand, and pebbles	A big piece of the hillside slid down. There is mud at the bottom.
Sandy beach	wind	very small—sand	
Mudflat	water	very small—silt	The sand has a ripple pattern.

SCIENCE AND ENGINEERING PRACTICES

Obtaining, evaluating, and communicating information

ELA CONNECTION

These suggested strategies address the Common Core State Standards for ELA.

W 8: Recall information from experiences or gather information from provided sources to answer a question.

SL 2: Recount or describe key ideas.

BREAKPOINT

23. Focus question: How does water and wind change landforms?

Remind students of the reading "Rocks Move." Ask the second focus question and project or write it on the board.

➤ *How does water and wind change landforms?*

Tell students that they have been investigating weathering, the breaking down of rock into smaller and smaller particles. They have used tools to separate the particles and observed their properties. Now they are going to learn more about what happens to those particles.

24. Introduce *erosion*

Refer students to the pictures in the article "Rocks Move" and ask them if they think the changes in the land happened quickly or slowly. Tell students that rocks can move quickly on a mudflat, across a beach, or in a flood.

Ask students to think about how long it might take for boulders to break off a mountain and for the boulder to break into smaller and smaller pieces—a process we call weathering. How long would it take for those pieces to break off and to move by wind and water to another location, and to end up at a beach by the ocean? Would it take a short time or a long time?

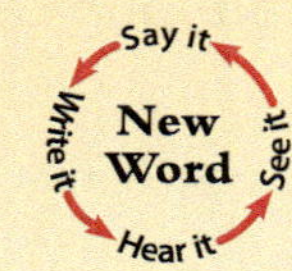

Introduce the word ***erosion*** to mean the movement of weathering rock to new locations.

25. View the video: All about Land Formations

Tell students the video will tell more about how the movement of rocks can change the land slowly to create new landforms. Show the 21-minute video, *All about Land Formations*. List new words on the word wall as they are discussed in the video.

At 12:26 minutes, there is a demonstration (Earth Inquires) of how landforms are created. It begins with these words:

"Moving water is a powerful force and can change the Earth's surface in a hurry. But some of the biggest changes in the Earth's surface are caused when moving water has a long time to work. Not even mountains can stand up to moving water wearing away at them for thousands of years."

SCIENCE AND ENGINEERING PRACTICES

Obtaining, evaluating, and communicating information

26. Discuss video demonstration

The video goes on to show two students developing a model of a mountain made of sand and gravel in a baking pan. Using a cup with holes, rain water falls on the mountain and erosion occurs. Stop the video BEFORE the water is poured on the model mountain so that you can ask students to discuss in their groups how this model compares to reality or an actual mountain. It will be important for students to distinguish between the model and the actual event and process that the model represents. Ask students to predict what might happen when the water is poured on the model mountain.

At the end of the demonstration, stop the video again and ask students to compare the model to the actual process. They should discuss scale both of the size of the mountain, the force of the water, and the time it would take for the changes to take place.

27. Discuss weathering and erosion

At the end of the video, distinguish between the breaking down of rocks into smaller pieces (weathering) and the movement or transport of those pieces to new locations to create new land formations (erosion) and the time it takes for these processes.

28. Conduct the demonstration live (optional)

If you have sand tables in the classroom or outdoors, you can conduct this demonstration live or have small groups carry it out. Determine what resources and facilities you have available for a live demonstration and decide how to proceed.

SCIENCE AND ENGINEERING PRACTICES

Developing and using models

TEACHING NOTE

Go to FOSSweb for Teacher Resources *and look for the Science and Engineering Practices—Grade 2 chapter for details on how to engage second graders with the practice of developing and using models.*

CROSSCUTTING CONCEPTS

Stability and change

READING *in Science Resources*

29. Read "Landforms"

Tell students that they will continue to explore different types of landforms in the reading "Landforms." Before reading, give students time to examine and discuss the photographs with a partner. Have them point out landforms they recognize from the videos and/or their own firsthand experiences.

Read the article together as a class. Call on individual students to read each of the captions aloud. Pause after each page and ask students to compare and contrast the different shapes and kinds of landforms.

Make a content grid on chart paper to help students find key details that support the main ideas in this text. See the sample below.

Reread the text aloud or have students read independently. Discuss at least one example from each page and as a class fill in the details on the chart. Add a simple drawing to represent each one.

Landform name	Illustration	How it was formed	Characteristics	Where it is found

Landforms

Some landforms are formed by eruption.

A **volcano** is a place where lava, ash, and **gases** escape from openings in Earth's crust.

A cinder cone is a kind of volcano. It forms when cinders (pieces of lava) burst out of Earth in an eruption.

24

ELA CONNECTION

These suggested strategies address the Common Core State Standards for ELA.

RI 2: Identify the main topic of the text.

RI 3: Describe the connection between scientific ideas or concepts.

RF 4: Read with accuracy and fluency to support comprehension.

The landforms photos include:

a. Landforms formed by eruptions

- Volcano (general photo)
- Cinder cone
- Composite
- Shield

b. Landforms formed by weathering and erosion

- Valley
- Canyon
- Mesa
- Butte
- Delta
- Beach
- Plain
- Sand dune

c. Raised surfaces of the land that can weather to make other landforms

- Mountain
- Plateau

30. Have a sense-making discussion

Ask students what landforms they can observe in the area. Add a few to the content grid chart.

Ask

➤ *How does wind change the land? Give some examples.*

➤ *How does water change the land? Give some examples.*

➤ *How does the land change over time?*

TEACHING NOTE

Refer to the Sense-Making Discussions for Three-Dimensional Learning chapter in Teacher Resources *on FOSSweb for more information about how to facilitate this with young students.*

31. Review vocabulary

Review key vocabulary added to the word wall in the video and reading about landforms using this strategy. Here are two suggestions for vocabulary review.

- Have students make a picture dictionary in their notebooks for each of these landforms.
- Give students a "deck" of landform word cards. In small groups, students take turns picking a card without showing the rest of the group and draw it on a whiteboard or scrap paper. The rest of the group tries to guess which landform the student drew.

beach
butte
canyon
delta
erosion
mesa
model
plain
plateau
sand dune
valley

32. Answer the second focus question

Tell students to label their drawings to identify the layers they see in the vial. Discuss what the layers should be called: sand, silt, and water. Ask them what they should label the space between the top of the water and the vial cap. [Air.]

Restate the focus question.

➤ *How does wind and water change landforms?*

Materials for Step 33

- *Paper plates*
- *Sand*
- *Clay*

33. Revisit the investigation guiding question

Give students a paper plate and some sand or clay. As a review of the investigation, ask students to do two things in their notebooks.

- Make a drawing that shows how a larger rock can become sand.
- Make a model of the landform you show them on a word card. When they finish their model, they should draw a picture in their notebook of the physical model they made. Use this for an embedded assessment.

SCIENCE AND ENGINEERING PRACTICES

Developing and using models

Constructing explanations

34. Assess progress: notebook entry

Collect student notebooks after class and check students' models.

What to Look For

- *Students draw pictures of the models they made of landforms.*
- *Students can name the landform models they drew.*

B R E A K P O I N T

35. Assess progress: I-Check

When students have completed the investigation, give them *Investigation 2 I-Check.*

Review student responses. Use the What to Look For information in the Assessment chapter for guidance. Note concepts that you might want to revisit with students, using the next-step suggestions.

The students' experiences in this investigation contribute to their understanding that some events on Earth happen quickly and others occur over a very long period of time, that wind and water can change the shape of the land, that water is found in many locations on Earth, and that matter can be described by its observable properties.

DISCIPLINARY CORE IDEAS

ESS1.C: The history of planet Earth

ESS2.A: Earth materials and systems

ESS2.B: Plate tectonics and large-scale system interactions

ESS2.C: The roles of water in Earth's surface processes

PS1.A: Structure and properties of matter

INTERDISCIPLINARY EXTENSIONS

Language Extensions

- **Write the journey of your rock**
  After reading "The Story of Sand," have students write their own stories of a rock they have found. Ask them to describe where they found a rock and how the rock came to be there.

- **Write rock stories**
  Ask students to choose one of the pictures in the "Rocks Move" article and write a story about it. Encourage students to be creative but scientifically accurate in their descriptions.

TEACHING NOTE

Refer to the teacher resources on FOSSweb for a list of appropriate trade books that relate to this module. These can be used to compare and contrast how the concepts of weathering and erosion are presented in two texts.

Math Extensions

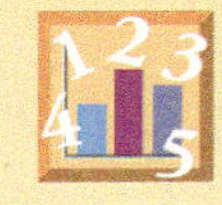

- **Math Problem A**
  Students determine how much money a particular collection of rocks will cost, based on the prices of a variety of rocks. If possible, set up a "rock shop" in your classroom with the rocks and the cost per rock.

  1. A girl wants to buy a river rock, three clay beads, and a mineral. How much money does she need?
  6¢ + 4¢ + 4¢ + 4¢ + 10¢ = 28¢

  2. A boy wants to buy three cobbles, two polished pebbles, and a piece of scoria. How much money does he need?
  5¢ + 5¢ + 5¢ + 7¢ + 7¢ + 8¢ = 37¢

  Notes on the problem. Have pennies, nickels, dimes, and quarters or plastic models of coins available as necessary to support students as they work on these problems. Remind students to show their work.

- **Math Problem B**
  Students are given money to spend in the rock shop to purchase rocks.

  1. You have 42¢ to spend at the rock shop. If you spend all your money, what rocks would you choose? Show your work.

  There are many combinations of rocks whose cost will add up to exactly 42 cents. Students may use known facts, such as 6 + 6 = 12 and 8 + 4 = 12, to determine which rocks to select.

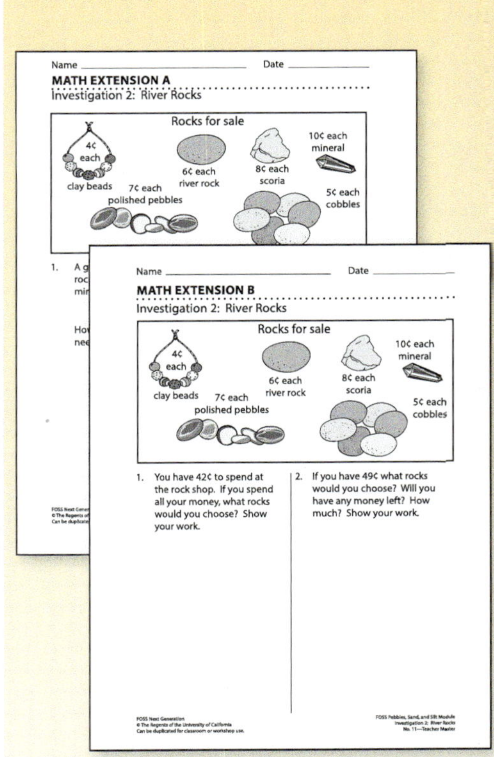

Name Date

MATH EXTENSION A

Investigation 2: River Rocks

Name Date

MATH EXTENSION B

Investigation 2: River Rocks

Rocks for sale

clay beads 4¢ each; polished pebbles 7¢ each; river rock 6¢ each; scoria 8¢ each; mineral 10¢ each; cobbles 5¢ each

1. You have 42¢ to spend at the rock shop. If you spend all your money, what rocks would you choose? Show your work.

2. If you have 49¢ what rocks would you choose? Will you have any money left? How much? Show your work.

Nos. 10–11—Teacher Masters

Two sample solutions:

3 minerals and 2 river rocks
(10¢ + 10¢ + 10¢ + 6¢ + 6¢)

2 cobbles, 2 minerals, 1 scoria, 1 clay bead
(5¢ + 5¢ + 10¢ + 10¢ + 8¢ + 4¢)

2. If you have 49¢ to buy rocks, what rocks would you choose? Will you have any money left? How much?

After they select their rocks, students determine how much, if any, change they will receive. Again, there are many possible solutions!

7 polished pebbles at 7¢ each
(49¢ exactly/no change)

4 minerals @ 10¢ each, 1 cobble @ 5¢
(45¢/3¢ change)

2 polished pebbles - 14¢, 2 scoria - 16¢, 3 river rocks 18¢
(48¢/1¢ change)

Notes on the problem. You can scale this problem depending upon the skills and abilities of your students. For example, students could be required to purchase at least three kinds of rocks or come up with two different choices of rocks to purchase, or purchase the greatest number of rocks with a given amount of money.

After students have an opportunity to solve problems independently, provide time to work with a partner. Have students present their solutions to the class.

Music Extension

- **Create rock music**
  Collect opaque, small, non-breakable containers with lids. Put gravel in one, sand in another, and pebbles in a third. Have students compare the sound that each makes when shaken.

Social Studies Extension

- **Visit a quarry**
  If there is a rock quarry or gravel pit in your area, make arrangements to take students there on a field trip. Help students find out about the different earth materials found in the quarry or pit, how they are removed, and how they are separated into different sizes for different uses.

  If an actual quarry is not available, search the Internet for a virtual tour of a rock quarry.

Science Extensions

- **Set up a screening station**
  Set out a basin of the river-rock mixture, a set of screens, and some 1/4 L containers. Ask students in pairs or groups of four to separate the mixture into different particle sizes. Provide vials and water if students call for them.

- **Make large river-rock shake-up bottles**
  Make one or more bottles like the one illustrated on notebook sheet 11, *Rocks in Bottle Drawing*. Add different combinations of rock material to some water. A recycled 1 or 2 L plastic soda or juice bottle works well.

TEACHING NOTE

Review the online activities for students on FOSSweb for module-specific science extensions.

- **Set up a sand exploration center**
  Get a large bus tray, a couple liters of sand, a variety of containers, spoons, sieves, and a container of water. Let pairs of students spend some time at this sand exploration center. They should try these activities.
  - Molding dry and wet sand
  - Pouring sand into and through various containers
  - Making sand castles
  - Using sieves

  Discuss the difference between dry and wet sand.
- **Look for clay soils**
  Tell students that clay soils are sticky like the clay they worked with. A mud ball made with clay soil will stay together when squeezed. Ask students to look for clay around the neighborhood. (You will use this information in Part 5 of Investigation 3.) Invite a soil conservationist to class to discuss clay soils.

Home/School Connection

At home, families reinforce the concept of size sequencing with a game of I Spy. In this version, players put five to ten objects on a table. They use objects that have a similar shape but vary in size.

Print or make copies of teacher master 12, *Home/School Connection* for Investigation 2, and send it home with students at the end of Part 2.

Another home/school connection activity is to have students talk with their families about landforms that are significant in their families' history. Is there a favorite place they like to visit? Are there famous landmarks to which they are connected? What landforms define their community?

HOME/SCHOOL CONNECTION
Investigation 2: River Rocks

Play I Spy. Gather five to ten objects that share a property and place them on a table. A set might include pencils, pens, flatware, straws, and chopsticks because they are all long and narrow. A set of books and catalogs might constitute a second set, a collection of stuffed toys a third, and so on.

Two players play the game. First the players organize the objects from smallest to largest. Then one player secretly chooses one object and compares it to the others: "I spy something that is bigger than ______ and smaller than ______." The second player guesses which object was chosen by player 1. If the guess is incorrect, player 1 provides a second "I spy" hint.

Swap roles and play again. Choose new sets of objects.

FOSS Pebbles, Sand, and Silt Module
Investigation 2: River Rocks
No. 12—Teacher Master

No. 12—Teacher Master

Investigation 3: Using Rocks

INVESTIGATION 3 — *Using Rocks*

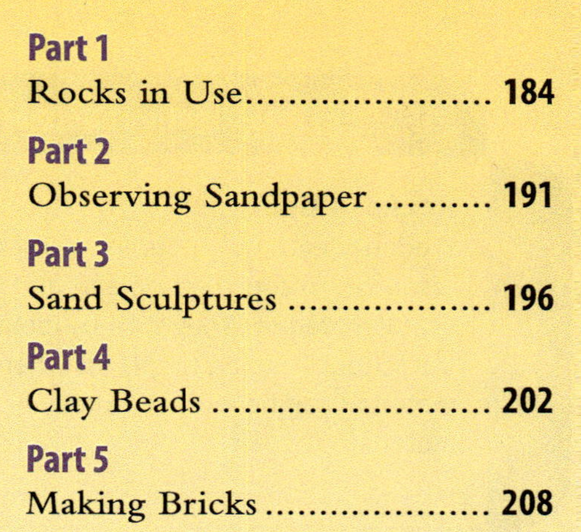

Guiding question for phenomenon:
How are different sizes of rock used as resources to make useful objects?

PURPOSE

Students investigate the ways that the properties of rocks of different sizes can be used to make useful objects. They discover that rock as a resource is a natural phenomenon occurring in predictable locations all over Earth's surface.

Content

- Earth materials are natural resources. The properties of different materials make them suitable for specific uses.
- Different sizes of sand are used on sandpaper to change the surface of wood from rough to smooth.
- Earth materials are commonly used in the construction of buildings and streets and used in sculptures and jewelry.

Practices

- Explore places where earth materials are naturally found and ways that earth materials are used. Search for earth materials outside the classroom.
- Use sand to make sculptures and clay to make beads, jewelry, and bricks.

Science and Engineering Practices

- Defining problems
- Planning and carrying out investigations
- Analyzing and interpreting data
- Constructing explanations
- Engaging in argument from evidence
- Obtaining, evaluating, and communicating information

Disciplinary Core Ideas

PS1: How can one explain the structure, properties, and interactions of matter?

PS1.A: Structure and properties of matter

ETS1: How do engineers solve problems?

ETS1.A: Defining and delimiting engineering problems

ETS1.B: Developing possible solutions

ETS1.C: Optimizing the design solution

Crosscutting Concepts

- Cause and effect
- Scale, proportion, and quantity
- Energy and matter

INVESTIGATION 3 — *Using Rocks*

	Investigation Summary	Time	Focus Question for Phenomenon, Practices
PART 1	**Rocks in Use** Students learn how people use rocks as natural resources to construct objects and to make useful materials. They start by looking outside the school building for places where earth materials can be found naturally or as building materials.	**Active Inv.** 1 Session * **Reading** 1 Session	**How do people use earth materials?** **Practices** Planning and carrying out investigations Analyzing and interpreting data Obtaining, evaluating, and communicating information
PART 2	**Observing Sandpaper** Students observe sandpaper and compare it to sand. They make and compare rubbings of three grades of sandpaper. Students compare the effectiveness of each grade of sandpaper in sanding a stick and make a claim from evidence.	**Active Inv.** 1 Session	**What does sand do for sandpaper?** **Practices** Defining problems Planning and carrying out investigations Analyzing and interpreting data Engaging in argument from evidence
PART 3	**Sand Sculptures** Students mix sand with a cornstarch matrix to make durable sand sculptures. They monitor the mixing process to determine the best amount of sand to mix with a given amount of matrix and analyze the results.	**Active Inv.** 1 Session	**How can we make a sand sculpture?** **Practices** Planning and carrying out investigations Analyzing and interpreting data Constructing explanations
PART 4	**Clay Beads** Students use clay to make beads or something decorative, which they paint and keep as a memento of their investigation of clay.	**Active Inv.** 2 Sessions	**What makes clay useful in making objects like beads?** **Practices** Constructing explanations
PART 5	**Making Bricks** Students make adobe clay bricks with a mixture of clay soil, dry grass or weeds, and water. After the bricks dry, they can be used to build a class wall.	**Active Inv.** 2 Sessions **Reading** 1 Session **Assessment** 1 Session	**How are bricks made?** **Practices** Constructing explanations Obtaining, evaluating, and communicating information

* A class session is 45–50 minutes.

Content Related to DCIs	Writing/Reading	Assessment
• Earth materials are natural resources. • The properties of different earth materials make them suitable for specific uses. • Earth materials are commonly used in the construction of buildings and streets.	**Science Notebook Entry** *Rocks in Use* **Science Resources Book** "Making Things with Rocks"	**Embedded Assessment** Science notebook entry
• The properties of different earth materials make them suitable for specific uses. • Different sizes of sand are used on sandpaper to change the surface of wood from rough to smooth.	**Science Notebook Entry** *Looking at Sandpaper*	**Embedded Assessment** Science notebook entry
• The properties of different earth materials make them suitable for specific uses. • Earth materials are used to make sculptures and jewelry.	**Science Notebook Entry** Answer the focus question	**Embedded Assessment** Performance assessment
• The properties of different earth materials make them suitable for specific uses. • Earth materials are used to make sculptures and jewelry.	**Science Notebook Entry** *Uses of Earth Materials* **Online Activity** "Find Earth Materials"	**Embedded Assessment** Science notebook entry
• The properties of different earth materials make them suitable for specific uses. • Simple bricks are made by combining clay soil with plant material.	**Science Notebook Entry** Answer the focus question **Science Resources Book** "What Are Natural Resources?"	**Benchmark Assessment** *Investigation 3 I-Check* **NGSS Performance Expectations addressed in this investigation** 2-PS1-1 2-PS1-2 K–2 ETS1-1; K–2 ETS1-2; K–2 ETS1-3

BACKGROUND *for the Teacher*

It's interesting to consider how much we depend on and take for granted the **natural resources** used so widely in construction. Walls are formed with **bricks** and **concrete** blocks, windows are made of glass, and roads and **sidewalks** are surfaced with cement or **asphalt** paving—and all of them use rocks. These materials are selected because their properties meet the builder's criteria.

Early humans might have been inspired to use earth materials for building by observing the creative industry of animals. In Africa the volcano-shaped mud homes of termites dot the landscape. Birds use mud as the matrix connecting the twigs and leaves in their nests. Beavers use mud for waterproofing their lodges and dams.

How Do People Use Earth Materials?

"Walls are formed with **bricks** *and* **concrete** *blocks, windows are made of glass, and roads and* **sidewalks** *are surfaced with cement or* **asphalt** *paving—and all of them use rocks."*

Here is a list of some of the other common earth materials used by **engineers** and others for construction, manufacturing, and landscaping.

Aggregate A mass of rock particles, such as pebbles, gravel, and sand.

Asphalt A dark-brown to black hydrocarbon-based material with a consistency varying from thick liquid to solid. Sand and gravel are added to asphalt to change its texture and strength.

Cement A finely ground rock powder (usually limestone with some additional minerals) that sets and **hardens** when mixed with water.

Concrete A mixture of cement, gravel, and/or sand that, when mixed with water, will set and harden. Stucco and **mortar** are special kinds of concrete: mortar is used between other earth materials, and stucco is applied as a surface layer.

Glass A mixture of quartz sand, lime, and soda, melted together and formed into containers, windows, and other objects.

Mortar Lime or cement or a combination of both, mixed with sand and water and used as a bonding agent between bricks or rocks.

Portland cement A cement consisting of a mixture of ground limestone and shale that has been heated until almost fused and then finely ground.

Potter's clay An iron-free, shapable clay especially suitable for making pottery, modeling, or throwing on a potter's wheel. Sand is often added to the clay to help the potter get a grip on the clay as it turns on the wheel. When clay is heated to a high temperature (fired), the particles fuse and no longer disintegrate in water.

What Does Sand Do for Sandpaper?

Sandpaper is heavy paper that has been coated with abrasive particles and is used as a tool, like a saw, chisel, or file. The particles on the sandpaper have rough edges that cut the wood (or metal or plastic or ceramic) surface. These rough, cutting edges can be used to remove small amounts of surface material, either to make it smoother or sometimes to make it rougher. The result depends largely on one property of the sandpaper, texture.

Close-up of sandpaper

Sandpaper was used in China in the thirteenth century. It was made of crushed shells, seeds, and sand glued with natural gum to parchment. Today there are many types of sandpaper for commercial and household uses, made with different types of paper or backing, a variety of particles, and a number of bonding techniques.

The "grit designation" of the sandpaper refers to the size of the particles embedded in the sandpaper. The lower the grit number, the larger the particle size and the coarser the paper. Sandpaper sheets are normally marked on the back with the numbered grit size and/or the grade. In this module we use #50 **coarse**, #80 **medium**, and #150 **fine** sandpapers.

Variety of kinds of sandpaper

Sandpaper was once made with quartz sand, but the materials used today are garnet (from the mineral almandine, which is harder than quartz; good as a final finish paper for wood), corundum or aluminum oxide (most common, long lasting, and good on all surfaces), silicon carbide (harder than garnet or aluminum oxide and better suited to metal, paint, or plastic than wood; good to use when wet), and emery (suited to fine polishing of metal).

How Can We Make a Sand Sculpture?

Sand is one of the most common materials on the earth's surface and comes from the physical and chemical weathering of rocks. The minerals in the parent rock partly determine the physical properties of the sand. Other factors contributing to its properties are the environmental conditions in the history of the sand—where it started, how far it traveled, and what happened to it over the years. A common sand found in deserts and on continental beaches is silicon dioxide derived from quartz, and aluminum silicate derived from feldspar. White, tropical beach sand is often calcium carbonate derived from limestone. Collecting sand from different locations around the world is a fascinating hobby because of the differences in color, texture, size, and other physical properties of the tiny bits of rock.

Different kinds of sand are used for different purposes. The construction aggregate industry is the largest consumer of sand for use in asphalt and concrete products. Another large consumer of sand is the glass industry. The sand used to make concrete may be quite different in physical and chemical composition from the sand used to make glass.

© iStockphoto/Sufi70
Sand up close

In construction, sand is used in stucco, plaster, roofing, bricks, grout, nonskid floors, paint, plastics, sealants, and rubber. In recreation, it is used for baseball diamonds and golf courses to improve drainage, for public playgrounds in volleyball courts and sandboxes, and in swimming-pool filters. Sand is mixed with salt to spread on icy roads for safety and traction. And it is used in commercial foundry applications to create cores and molds in the casting of iron, steel, copper, and aluminum products. Foundry sand is the second largest industrial use of sand in terms of tons consumed.

Not only do we rely on sand for the production of essential products, but we enjoy resting on it and playing in it. Making sand castles is a favorite way to relax and enjoy sand. These temporary structures are held together with water and compaction. In this investigation, students will add a **matrix** to the sand and create **sculptures** that are more durable.

What Makes Clay Useful in Making Objects Like Beads?

Clay, an earth material that has a particle size smaller than sand, has been used since prehistoric times to make vessels for storing liquids. Clays mixed with water have a plasticity that allows them to be formed into shapes. When dry, the clay changes and becomes firm and hard. And if the clay is then heated (fired) in a kiln, permanent physical and chemical changes take place, converting clay into a ceramic material. Clay is important in the production of porcelain, china, and earthenware, as well as tile for wall and floor coverings, and pipes for drainage and sewage. Clay can be used to make pottery objects that are useful or just decorative, such as beads.

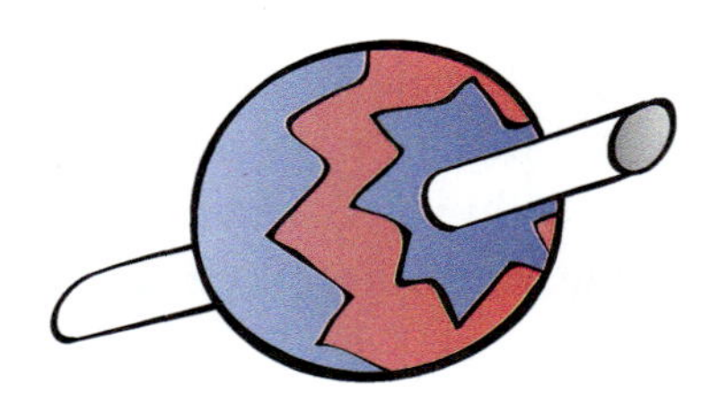

How Are Bricks Made?

Molded mud bricks have been used for nearly 8,000 years. Various traditional cultures still use soil and mud to **build** their homes. They make a mixture of straw, earth, and water called adobe. When adobe is wet, it can be formed into bricks. Molded adobe bricks placed in the sun bake hard and can be used for construction. Students follow an adobe recipe to make bricks in this investigation.

The characteristically red bricks seen in walls and buildings are made using the same basic process as adobe with the addition of a firing step, which creates a harder, stronger brick. Bricks begin with a rock called shale (composed of hardened clay), which is ground into a fine powder, and mixed with water until it's the consistency of toothpaste. The clay mixture is pushed through a tube into rectangular forms. Metal rods in the tubes make holes in the slugs of shaped clay. The rectangular slugs are cut into bricks and dried slowly (60 hours) so they don't crack. Firing then happens in kilns at temperatures nearing 1200°C (2192°F). The bricks are cooled slowly and stacked on pallets to be shipped to construction sites.

TEACHING CHILDREN *about Using Rocks*

NGSS Foundation Box for DCI

PS1.A: Structure and properties of matter

- Different kinds of matter exist and many of them can be either solid or liquid, depending on temperature. Matter can be described and classified by its observable properties. (2-PS1-1)
- Different properties are suited to different purposes. (2-PS1-2)

ETS1.A: Defining and delimiting engineering problems

- Before beginning to design a solution, it is important to clearly understand the problem. (K-2-ETS1-1)

ETS1.B: Developing possible solutions

- Designs can be conveyed through sketches, drawings, or physical models. These representations are useful in communicating ideas for a problem's solutions to other people. (K-2-ETS1-2)

ETS1.C: Optimizing the design solution

- Because there is always more than one possible solution to a problem, it is useful to compare and test designs. (K-2-ETS1-3)

Developing Disciplinary Core Ideas (DCI)

To primary students, things are what they are. Glass is glass, brick is brick, concrete is concrete, and on and on. And why not? Most of us are isolated from the processes that produce these materials. We see only the end product.

One of the roles of science and engineering education is to help students take their world apart and see what it is made of. This module uses this analytical approach to understanding. Simple traditional materials like brick and clay become totally comprehensible when students analyze them to discover their composition, and students achieve even greater knowledge of the material world when they reconstruct objects like bricks, using everyday processes.

A simple field trip around the school grounds or into the immediate neighborhood can be a wonderful experience for students *and* the adults guiding them. Primary students see the world in interesting ways. Some have a hard time concentrating on a task such as looking for sand, gravel, and pebbles in use in the environment; they are more concerned with who is first in line, checking to see who else is in the yard, watching traffic on the street, or just running around. Such expansive students will need constant focusing if they are to benefit from a specific task set for the outdoor tour.

Other students will be completely engrossed in the discovery activity of finding earth materials in use. They will search with enthusiasm, and each find will be a victory.

As always, it is not the accuracy with which students analyze the contents of a building material that is important, but rather the processes of observing closely, comparing what they see with what they learned earlier, and communicating their discoveries. That's science. In addition, we are asking students to apply their knowledge in the service of society—how can we make useful objects from earth materials and at the same time conserve the resources we use. That's engineering!

By working with a variety of earth materials and using those materials to produce a variety of objects, students come to understand that different properties are suited to different purposes, and this science knowledge makes it possible to engineer successful objects. The experiences students have in this investigation contribute to the disciplinary core ideas **PS1.A, Structures, and properties of matter,** and the core ideas of **engineering design** (ETS1.A, ETS1.B, and ETS1.C).

Engaging in Science and Engineering Practices (SEP)

In this investigation, students engage in these practices.

- **Defining problems** related to the development of sandpaper to explore what engineers were trying to solve.
- **Planning and carrying out investigations** by making observations and collecting data about the use of earth materials to make everyday things. Students use and share with partners pictures, drawing, and written observations recorded in their notebooks.
- **Analyzing and interpreting data** from tests with sandpaper and sand sculptures to determine if the product and tool work as intended.
- **Constructing explanations** by making firsthand observations to construct an evidence-based account for how clay is a good material for making beads or for making bricks. Students compare the effect of different kinds of sandpaper (the solution or tool) on the sanding of wood and the results of different mixtures of sand and matrix on making sand sculptures.
- **Engaging in argument from evidence** involving the effect of different kinds of sandpaper on wood.
- **Obtaining, evaluating, and communicating information** about the use of earth materials by engineers to make things for everyday life. Students gather information from text to answer questions, and communicate information orally and in written forms using drawings, words, and numbers.

NGSS Foundation Box for SEP

- **Define a simple problem** that can be solved through the development of a new or improved object or tool.
- **Make observations** (firsthand) to collect data that can be used to make comparisons.
- **Record information** (observations, thoughts, and ideas).
- **Use and share pictures, drawings,** and/or writings of observations.
- **Analyze data from tests** of an object or tool to determine if it works as intended.
- **Make observations** (firsthand or from media) to construct an evidence-based account for natural phenomena.
- **Generate and/or compare solutions** to a problem.
- **Construct an argument** with evidence to support a claim.
- **Read grade-appropriate text** and/or use media to obtain scientific and/or technical information to determine patterns in and/or evidence about the natural and designed world(s).
- **Obtain information** using various texts, text features, and other media that will be useful in answering a scientific question.
- **Communicate information** or design ideas and/or solutions with others in oral and/or written forms using models, drawings, writing, or numbers that provide detail about scientific ideas, practices, and/or design ideas.

INVESTIGATION 3 — Using Rocks

Exposing Crosscutting Concepts (CC)

In this investigation, the focus is on these crosscutting concepts.

- **Cause and effect.** Sanding wood with different grades of sandpaper produced different effects based on the size of the sand on the paper; the results of mixing sand with matrix to make sculptures depends on the proportions of the two materials that you mix.
- **Scale, proportion, and quantity.** Mixing different proportions of sand and matrix results in different outcomes. To create a useful product, the proportions need to be specified.
- **Energy and matter.** Different sizes and amounts of earth materials go together to create bricks, a new product.

NGSS Foundation Box for CC

- **Cause and effect:** Events have causes that generate observable patterns; simple tests can be designed to gather evidence to support or refute student ideas about causes.
- **Scale, proportion, and quantity:** Relative scales allow objects and events to be compared and described (bigger and smaller; hotter and colder; faster and slower).
- **Energy and matter:** Objects may break into smaller pieces, be put together into larger pieces, or change shapes.

Connections to Science, Technology, Society, and the Environment

- **Influence of engineering, technology, and science on society and the natural world.** Every human-made product is designed by applying some knowledge of the natural world and is built by using natural materials.

Connections to the Nature of Science

- **Science addresses questions about the natural and material world.** Scientists study the natural and material world.

Asphalt
Brick
Build
Coarse
Concrete
Engineer
Fine
Harden
Matrix
Medium
Mortar
Natural resources
Sandpaper
Sculpture
Sidewalk

Conceptual Flow

Rock as a natural resource is the phenomenon students investigate. The guiding question is how are different sizes of rock used as resources to make useful objects?

The **conceptual flow** for this third investigation starts with students observing **rocks as natural resources** in use in the schoolyard. In Part 1, students go outdoors to search for cobbles, **pebbles, gravel, and sand** as single earth materials or in a mixture serving a function. They are guided by a list of objects that they might find, such as clay flowerpots, concrete steps, a gravel or cobblestone path, a brick wall, mortar between bricks, garden landscaping rocks, or a sport field with sand. Students learn that **concrete** is a mixture of cement, gravel, sand, and **water**, as is mortar. The use of these materials depends on the properties of the earth materials and their availability. Earth materials are essential for a variety of important building products in our everyday life.

In Part 2, students continue to study the uses of earth materials by focusing on **sandpaper** and making **texture** rubbings of coarse, medium, and fine sandpapers. The size of the particle results in the different textures that are useful in repairing and finishing wood products.

In Part 3, students study the properties of **sand** by transforming it into **sculptures** using a matrix made of cornstarch and water. Sand can be temporarily formed by mixing and compacting it with water, but turning it into more durable sculptures requires a binder.

In Part 4, students work with **clay** and learn how to mold and shape this earth material to make a decorative bead. This experience adds to their understanding of clay developed in the previous investigation when they mixed it with water and let the particles separate and then settle.

In Part 5, students use local clay to make **bricks**. The local clay will have different properties than the clay used to make the decorative piece. By working and molding the clay into a brick and seeing what it is like when it dries, students will begin to understand the use of earth materials in construction.

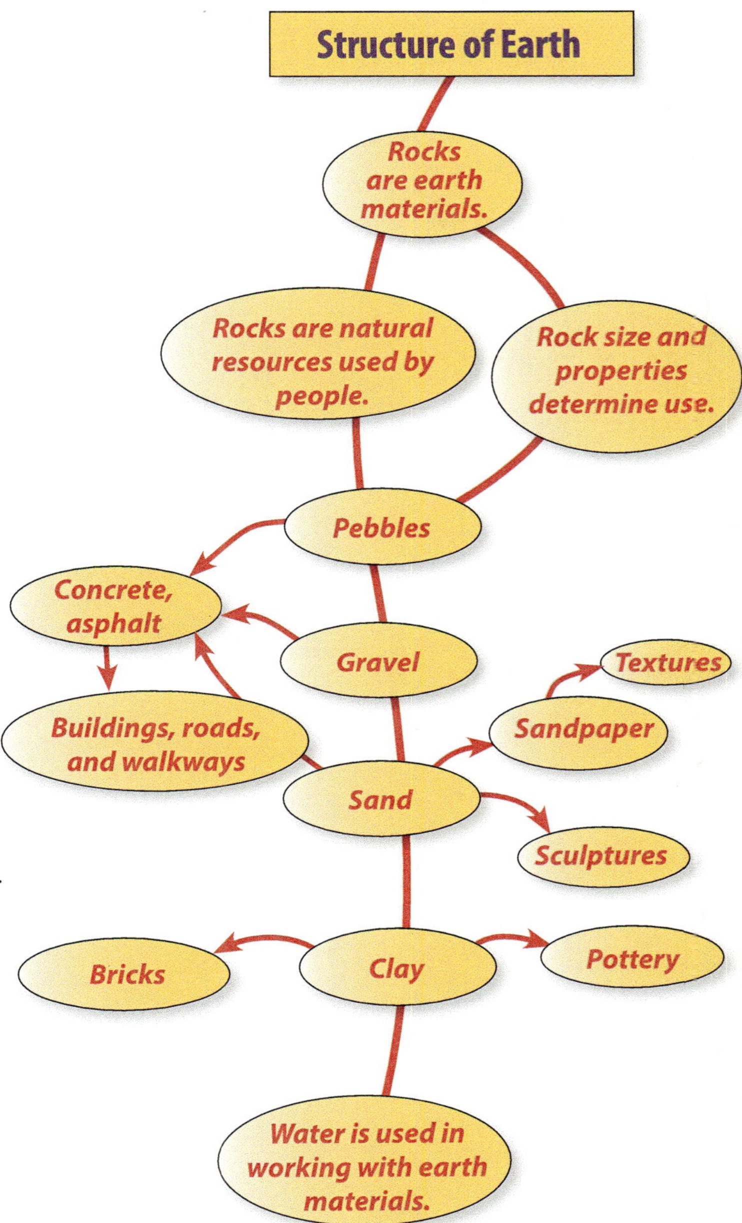

INVESTIGATION 3 — *Using Rocks*

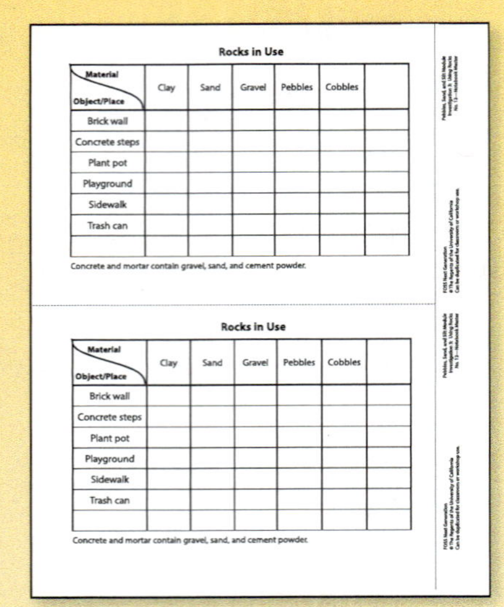

Rocks in Use

Material / Object/Place	Clay	Sand	Gravel	Pebbles	Cobbles	
Brick wall						
Concrete steps						
Plant pot						
Playground						
Sidewalk						
Trash can						

Concrete and mortar contain gravel, sand, and cement powder.

Rocks in Use

Material / Object/Place	Clay	Sand	Gravel	Pebbles	Cobbles	
Brick wall						
Concrete steps						
Plant pot						
Playground						
Sidewalk						
Trash can						

Concrete and mortar contain gravel, sand, and cement powder.

No. 12—Notebook Master

MATERIALS *for*

Part 1: *Rocks in Use*

For each student

- 1 Clipboard ★
- ❑ 1 Notebook sheet 12, *Rocks in Use*
- 1 *FOSS Science Resources: Pebbles, Sand, and Silt*
  - "Making Things with Rocks"

For the class

- 1 Camera (optional) ★
- 1 Big book, *FOSS Science Resources: Pebbles, Sand, and Silt*

For embedded assessment

- ❑ • *Embedded Assessment Notes*

★ Supplied by the teacher.   ❑ Use the duplication master to make copies.

GETTING READY *for*

Part 1: *Rocks in Use*

1. Schedule the investigation

This part will take one outdoor session for active investigation and one session for the reading.

TEACHING NOTE

You can do this outdoor activity out of order if the weather does not cooperate on the day you planned it.

2. Preview Part 1

Students learn how people use rocks as natural resources to construct objects and to make useful materials. They start by looking outside the school building for places where earth materials can be found naturally or as building materials. The focus question is **How do people use earth materials?**

NOTE

To prepare for this investigation, view the teacher preparation video on FOSSweb.

3. Enlist additional adults

Another adult or an older student will make this excursion richer for students. Divide the class into groups of eight to ten students. Have each group follow the same route. One or more adults could carry a camera for taking pictures of what students find.

4. Select outdoor site

Scout the schoolyard ahead of time to locate interesting sites. Look for brick and mortar walls, concrete blocks, concrete sidewalks, asphalt or blacktop, sculptures, clay pots, stones set in walls for decoration, garden areas, and so forth. Also look for evidence that rocks have been moved, such as gravel or sand around a downspout. Plan a walking route that will pass by the things you discovered. Check the route the morning of the activity for any problems.

5. Prepare clipboards for outdoor recording

As students find rocks in use, they should record their observations on the *Rocks in Use* sheet. Review the table on the sheet and plan how to introduce it to students in Step 3 of Guiding the Investigation.

TEACHING NOTE

See the Taking FOSS Outdoors chapter for suggestions on outdoor recording.

Students will inevitably find a rock, a piece of concrete or brick, or some other treasure they want to bring to the classroom. Decide ahead of time how to handle this. You could tell students that you are making a memory collection, so no collections.

6. Plan to read *Science Resources*: "Making Things with Rocks"

Plan to read "Making Things with Rocks" during a reading period after completing this part.

7. Plan assessment: notebook entry

In Step 9, students answer the focus question and glue the *Rocks in Use* sheet into their notebooks. Check that students are able to match earth materials with their uses.

FOCUS QUESTION

How do people use earth materials?

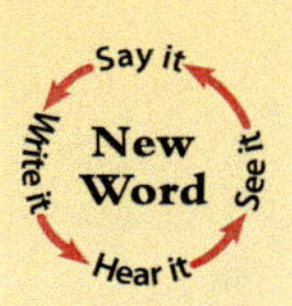

GUIDING *the Investigation*

Part 1: *Rocks in Use*

1. Review particle sizes

Gather students at the rug. Ask them to tell you the different sizes of rocks they have looked at in this module. [Pebbles, gravel, sand, silt or clay.] Write the names on the board in order.

2. Focus question: How do people use earth materials?

Ask the focus question and project or write it on the board.

➤ *How do people use earth materials?*

Brainstorm ideas and write them on the board. Spend a few minutes eliciting student ideas.

3. Describe the outdoor field trip

Tell students,

*Some scientists are **engineers**. Engineers use what they know about the properties of materials to **build** useful things.*

Many human-made products are designed by engineers using information about the properties of natural materials such as those we have been studying—earth materials.

We are going on a field trip today. We are going around the schoolyard to look for as many places as possible where we can find earth materials in use. We'll keep notes on this sheet about what we find as we go.

Hold up a clipboard with notebook sheet 12, *Rocks in Use*, on it. Point out the names of the earth materials across the top of the table and the names of objects or places along the left side of the table.

Rocks in Use

Material / Object/Place	Clay	Sand	Gravel	Pebbles	Cobbles	
Brick wall	X	X				
Concrete steps						
Plant pot						
Playground						
Sidewalk						
Trash can						

Concrete and mortar contain gravel, sand, and cement powder.

Distribute a clipboard and pencil to each student. Together read the names of the earth materials across the top of the table and the objects or places down the side. Model how to mark *X*s at the intersection between an object and all of the earth materials they can observe in that object. For example, if they observe a brick outdoors, they might find that it is made of clay and sand.

4. Go outdoors

Have students line up in buddy pairs. Have each student take their clipboard and pencil and walk outdoors, using your outdoor learning door. Form a sharing circle and describe where the search will be conducted and how much time students will have to search for earth materials in use. You might get students started by showing them one location where earth materials are in use.

Circulate to the groups and ask questions such as,

➤ *Where did you find sand in use? Why is sand used in the **sidewalk**?*

➤ *Where did you find pebbles and gravel? Why are pebbles and gravel used in playground **asphalt**?*

➤ *Where did you find **bricks** and what were they made of? Did the bricks look the same?*

➤ *What are the curbs and steps made of?*

➤ *Where did you find **concrete**? Did all the concrete look the same?*

5. Introduce vocabulary

As you circulate to the groups, take the opportunity to introduce or reinforce words while students are exploring a particular material. Some new words that might come up are concrete, **mortar**, and asphalt or blacktop. This is a good time to explain that concrete is a mixture of gravel, sand, and cement powder. Make the distinction between cement (what goes into the mixture) and concrete, the result of the mixture of gravel, sand, cement, and water. Mortar is another mixture of gravel, sand, and cement. It is used to hold brick or concrete blocks together.

6. Share observations

Give your signal for students to return to the sharing circle with their clipboards. Restate the focus question.

➤ *How do people use earth materials?*

Go around the circle and ask a few students to describe one location where they found earth materials in use. If other students found that same use, ask them to raise their hands.

SCIENCE AND ENGINEERING PRACTICES

Planning and carrying out investigations

Analyzing and interpreting data

Materials for Step 4

- *Clipboards*
- ***Rocks in Use** sheets*
- *Camera (optional)*

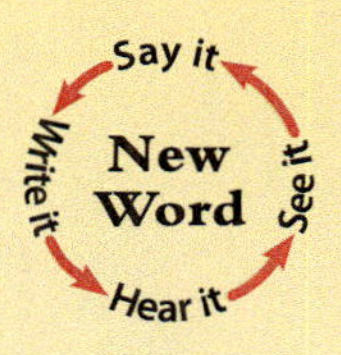

SCIENCE AND ENGINEERING PRACTICES

Analyzing and interpreting data

7. Return to class

Return to the classroom. Go through the list of sizes on the board one by one, and have students describe locations where they found each material. For example, under "pebbles" they might list sidewalks and the blacktop of the playground. Go through the whole list, recording all the places where students discovered the materials.

asphalt
brick
build
concrete
engineer
mortar
sidewalk

8. Review vocabulary

Review the vocabulary introduced earlier and, if necessary, add words to the word wall.

Make sets of word cards for each pair of students using the new vocabulary words plus clay, sand, pebbles, and gravel. Have students sort the words into two groups—earth materials and uses. Next, have students take a word card from "uses" and match it with the earth materials cards used to make that material. Have them repeat the process for each of the "uses." Encourage students to use their notebooks and the class word wall as a reference.

9. Answer the focus question

Have students glue notebook sheet 12, *Rocks in Use*, into their notebooks. Have them write and answer the focus question on the next blank page.

➤ *How do people use earth materials?*

10. Assess progress: notebook entry

Check students' focus question answers and *Rocks in Use* notebook sheets.

What to Look For

- *Based on their field trip observations, students are able to describe how people use earth materials.*
- *Students are able to organize data by filling in the chart, marking appropriate earth material sizes in objects they found made of earth materials.*

READING *in Science Resources*

11. Read "Making Things with Rocks"

Ask students what is made with rocks. Brainstorm ideas, referring to the schoolyard field trip to stimulate ideas. Explain that "Making Things with Rocks" will tell about other things that are made with rocks.

Let students preview the text by looking at and discussing what they observe in the photographs. Remind students that one way to help them understand text is to take notes.

12. Use the note making strategy

Give each student three or more self-stick notes to mark the pages using symbols where they learned something new (L), a question they have (?), and any confusing words or phrases (C). See the Science-Centered Language Development Chapter for a list of other symbols students can use to annotate the text.

Have students read independently. For emergent readers, pull a group aside and read aloud or use guided reading techniques.

13. Have a sense-making discussion

Tell students to share their notes with their table group and to help each other figure out the meaning of any confusing words of phrases. Have students share a few questions for possible discussion topics later on. Review any confusing terms.

Use the following questions to check for student understanding.

- *What was made of rock in the article?* [Buildings, pavement, pots.]
- *What size rock was used for things made in the reading?* [All sizes.]
- *Where do people get rocks to make things?* [A quarry.]

Read the text aloud modeling strategies that support reading comprehension (asking questions, connecting to prior knowledge, visualizing, summarizing, etc.). Pose a few of the following questions for students to discuss in their groups and then share with the whole class. Encourage students to support their responses with evidence from the text.

- *How does the size of a rock affect its use?* [Big rocks are used for making big things; small particles like clay are used for making pottery and bricks.]
- *Why are rocks used in concrete?* [Rocks like gravel are part of the mixture to make pavement hard and strong. The cement is like glue holding the gravel together.]

Making Things with Rocks

People use rocks to make things. A quarry is a place where people dig rocks out of the ground.

31

SCIENCE AND ENGINEERING PRACTICES

Obtaining, evaluating, and communicating information

ELA CONNECTION

These suggested strategies address the Common Core State Standards for ELA.

RF 4: Read with accuracy and fluency to support comprehension.

RI 1: Ask and answer questions to demonstrate understanding.

RI 3: Describe the connection between scientific ideas or concepts.

TEACHING NOTE

When using rock of any size to build something, you need a matrix to hold the pieces together. Gravel in concrete sidewalks is held with cement; large stones in buildings are held with mortar (which is concrete); and particles of clay are bonded with moisture.

- *What size rock comes from a quarry?* [All sizes.]
- *Why are there all sizes of rock in a quarry?* [Large rocks are broken into many different smaller sizes, and they are all found in a quarry.]
- *How do the big rocks used in the church stay together and in place?* [With mortar.]
- *Why are people able to shape and mold clay? How does clay stay together?* [The particles of clay are very, very small. When the particles are mixed with water, they stay together and are easy to mold. When they dry, they harden in that position.]

WRAP-UP/WARM-UP

14. Share notebook entries

Conclude Part 1 or start Part 2 by having students share notebook entries. Ask students to open their science notebooks to the most recent entry. Read the focus question together.

- *How do people use earth materials?*

Ask student pairs to describe what they found in the schoolyard. Have them think about why the earth materials they saw in the schoolyard look different than the ones they see in pictures or observed in class. At this point, the focus is to get them to consider the purpose of the objects they observed. They will continue to consider this over the course of this investigation.

Finally, ask students what questions they have about humans using earth materials.

TEACHING NOTE

See the ***Home/School Connection*** *for Investigation 3 at the end of the Interdisciplinary Extensions section. This is a good time to send it home with students.*

MATERIALS *for*

Part 2: *Observing Sandpaper*

For each student

- 1 Piece of fine sandpaper, 8 × 9 cm (3" × 3 5/8")
- 1 Piece of medium sandpaper, 8 × 9 cm (3" × 3 5/8")
- 1 Piece of coarse sandpaper, 8 × 9 cm (3" × 3 5/8")
- 1 Half sheet of white paper ★
- 1 Hand lens
- 1 Craft stick
- ❑ 1 Notebook sheet 13, *Looking at Sandpaper*

For the class

- 1 Vial
- 2 Loupes/magnifying lenses
- 2 Paper plates
- • Sand, clean
- 8 Zip bags, 1 L
- • Pencils or crayons ★

For embedded assessment

- ❑ • *Embedded Assessment Notes*

★ Supplied by the teacher.   ❑ Use the duplication master to make copies.

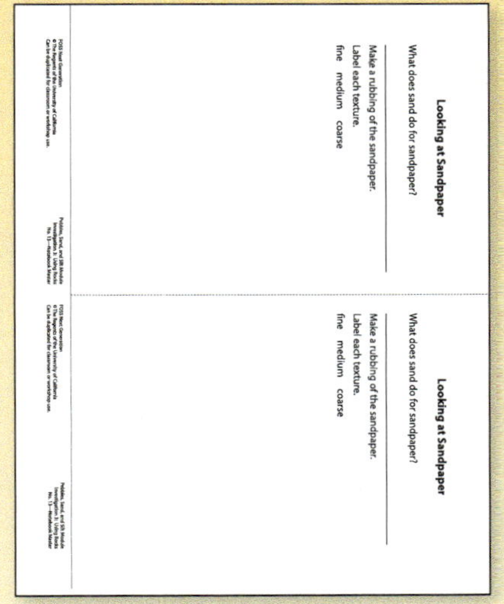

Looking at Sandpaper

What does sand do for sandpaper?

Make a rubbing of the sandpaper.
Label each texture.
fine medium coarse

Looking at Sandpaper

What does sand do for sandpaper?

Make a rubbing of the sandpaper.
Label each texture.
fine medium coarse

No. 13—Notebook Master

GETTING READY *for*

Part 2: *Observing Sandpaper*

1. Schedule the investigation

This part will take one active investigation session.

2. Preview Part 2

Students observe sandpaper and compare it to sand. They make and compare rubbings of three grades of sandpaper. Students compare the effectiveness of each grade of sandpaper in sanding a stick and make a claim from evidence. The focus question is **What does sand do for sandpaper?**

3. Prepare sandpaper pieces and practice sheets

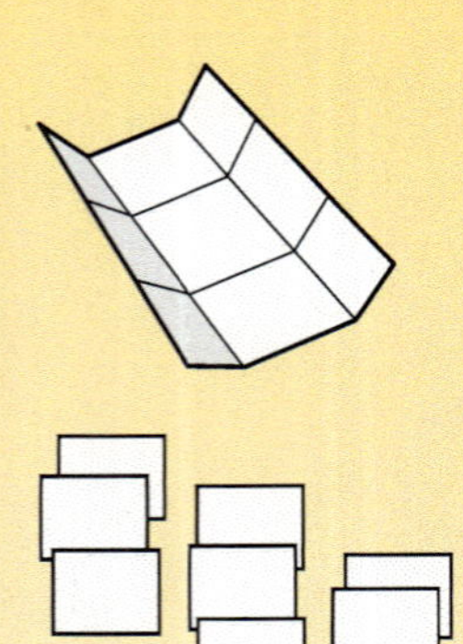

Tear the sandpaper into pieces before conducting this investigation for the first time. Sandpaper is not a consumable item and pieces will be reused many times. Fold each sheet of sandpaper in thirds, and crease each of the folds sharply, both forward and backward, so sheets tear cleanly into three strips. Fold, crease, and tear each strip into three equal parts.

Cut white paper in half. Each student will need one half sheet to practice making rubbings. Final rubbings will be done on notebook sheet 13, *Looking at Sandpaper.*

4. Prepare sand

Put three-quarters of a vial of sand in each of eight zip bags, one for each group. Students will use hand lenses and magnifying lenses/loupes to look at the sand through the bags.

Pour half a vial of sand onto two paper plates for direct viewing with the loupes/magnifying lenses.

5. Plan assessment: notebook entry

In Step 6, students complete notebook sheet 13, *Looking at Sandpaper.* Check that students understand the relationship between texture and size of sand on the sandpaper. Larger sand pieces means rougher texture.

GUIDING *the Investigation*

Part 2: *Observing Sandpaper*

FOCUS QUESTION

What does sand do for sandpaper?

Materials for Steps 1–2

- *Hand lenses*
- *Coarse sandpaper pieces*

Materials for Step 3

- *White paper*
- *Pencils or crayons*

1. Introduce the sandpaper

Explain to students that when humans use earth materials, they think about the properties of the resource. Tell students that you have a piece of paper for them to observe. Ask them to use a hand lens to find out what's on the paper. Don't tell them it is sandpaper at this time; let them find that out on their own.

2. Observe magnified sandpaper

Pass out one piece of coarse sandpaper to each student. Have students rub their fingers gently over the paper and use the hand lens to get a closer view. Ask,

➤ *What is on the paper?*

When they identify it as sand, tell them that sand has an important use that they will look at today: **sandpaper**.

3. Practice making sandpaper rubbings

Give each student a half sheet of paper and a pencil or crayon. Show them how to place the paper over the sandpaper, hold it securely in place, and rub with the side of a pencil to record the sandpaper texture.

If students have problems using pencils to create sandpaper rubbings, have them try peeled crayons.

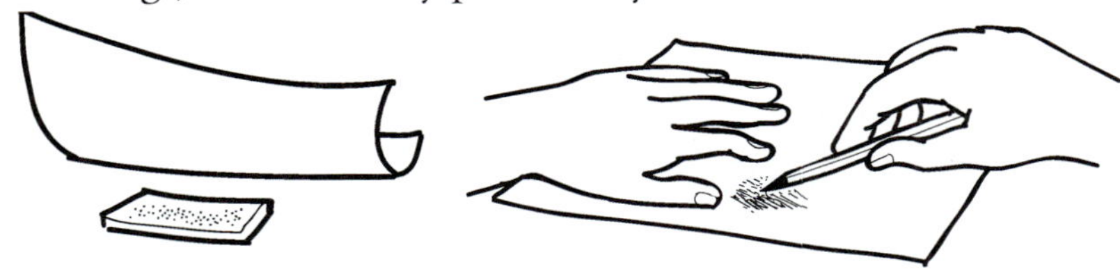

4. Introduce different sandpapers

Once students have their rubbing technique down, distribute samples of the medium and fine grit sandpaper. Have students rub their fingers gently over the new sandpapers and use the hand lens to get a closer view. Ask them what is different about each paper. Prompt students to describe the size of the sand and how that affects the texture of the paper. [The sand is a different size on each of the different papers; there are different numbers of particles on each paper; the texture of the paper is different.]

5. Introduce vocabulary

Tell students the sandpaper with the largest pieces of sand is **coarse** sandpaper, the paper with the middle-sized sand is the **medium** sandpaper, and the paper with the smallest sand is the **fine** sandpaper.

Ask students what sandpaper is used for. If they don't know, explain that it is used to make rough things, such as wood, smooth or to remove paint or other things from a surface.

6. **Focus question: What does sand do for sandpaper?**
Ask the focus question and project or write it on the board.

➤ *What does sand do for sandpaper?*

SCIENCE AND ENGINEERING PRACTICES

Planning and carrying out investigations

Distribute notebook sheet 13, *Looking at Sandpaper*, and review the text with students. Have them make a rubbing of each kind of sandpaper, compare the rubbings, and label them by writing the descriptive texture word on the paper next to each of the rubbings. The rubbings show what sand does for paper—it gives it texture.

7. **Identify sandpaper**
Have students challenge each other to identify the type of sandpaper with their eyes closed. Encourage them to use the new vocabulary.

8. **Rub sandpaper together**
Have students gently rub two pieces of sandpaper together to see what comes off. They can catch the sand on their pieces of white practice paper, and look at it with a hand lens.

Materials for Step 9–10

- *Sand in zip bags*
- *Sand on paper plates*
- *Loupes/magnifying lenses*
- *Craft sticks*

9. **Compare sand**
Distribute the samples of sand in the zip bags, one to each group and have students compare the sandpaper sand to these sand samples. Show students the sand samples on the paper plates at the learning center and the loupes/magnifying lenses students can use to observe the sand.

SCIENCE AND ENGINEERING PRACTICES

Analyzing and interpreting data

Engaging in arugument from evidence

10. **Have a sense-making discussion**
Give each student a craft stick and paper plate. Ask them to first rub the stick in the sand to see the results. Then have them use each kind of sandpaper to sand the stick. Ask them to compare the results of sanding using the different textures of sandpaper. Ask students to make a claim about the effect of using different kinds of sandpaper on a piece of wood and support their claim with evidence.

11. **Clean up**
Have students stack the three kinds of sandpaper in separate piles. The sandpaper in the kits is not a consumable item and will be reused many times.

12. **Review vocabulary**
Review key vocabulary added to the word wall during this part.

coarse
fine
medium
sandpaper

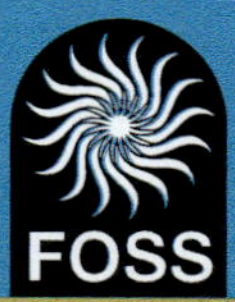

13. Answer the focus question

Have students glue notebook sheet 13, *Looking at Sandpaper*, into their notebooks. Have them write and answer the focus question on the next blank page.

➤ *What does sand do for sandpaper?*

Encourage students to think about how the size of the sand particles changes the texture of the sandpaper. They can use cause-and-effect statements to explain. For students who need support, you might provide sentence frames such as The bigger the pieces of sand, the _____ the texture. The texture of the sandpaper affects the _____ .

14. Assess progress: notebook entry

Collect students notebooks after class and check the *Looking at Sandpaper* notebook sheet. Check to see that students are relating the size of sand pieces on sandpaper to the texture.

What to Look For

- *Students analyzed their rubbings to explain that larger sand pieces create a rougher texture.*

CROSSCUTTING CONCEPTS

Cause and effect

WRAP-UP/WARM-UP

15. Share notebook entries

Conclude Part 2 or start Part 3 by having students share notebook entries. Ask students to open their science notebooks to the most recent entry. Read the focus question together.

➤ *What does sand do for sandpaper?*

Ask students to pair up with a partner to

- share their answers to the focus question;
- describe their labeled rubbings.

ELA CONNECTION

This suggested strategy addresses the Common Core State Standards for ELA.

SL 1: Participate in collaborative conversations.

Remind students of the engineering design process discussed earlier about the use of screens to separate rocks into groups of different sizes. Ask students to discuss what problems they think sandpaper was designed to solve. Have them brainstorm other ways these problems could be solved. Record student ideas for future projects. Tell students that in the next part they will explore another way to use sand to change or create something new.

SCIENCE AND ENGINEERING PRACTICES

Defining problems

MATERIALS *for*

Part 3: *Sand Sculptures*

For each pair of students

- 2 Paper plates
- 1 Cup of sand (half-full)
- 1 Metal spoon

For the class

- 2 Basins
- 4 Vials
- 1 Paper plate
- 1 Whisk broom and dustpan
- • Clean sand, 1 L, about 2.3 kg (5 lb.)
- 1 Box of cornstarch, 454 g (1 lb.) ★
- • Water ★
- 1 Saucepan ★
- 1 Mixing spoon ★
- 1 Container or jar, with lid, or zip bag, about 1 L ★
- • White glue ★
- 1 Piece of landscaping sandstone (optional) ★

For embedded assessment

❑ • *Performance Assessment Checklist*

★ Supplied by the teacher.

❑ Use the duplication master to make copies.

GETTING READY *for*

Part 3: *Sand Sculptures*

1. Schedule the investigation

This part will take one session for the whole class to make and discuss sand sculptures. The activity could also be done at a center, which will require 15–20 minutes for each group.

2. Preview Part 3

Students mix sand with a cornstarch matrix to make durable sand sculptures. They monitor the mixing process to determine the best amount of sand to mix with a given amount of matrix and analyze the results. The focus question is **How can we make a sand sculpture?**

3. Make sand matrix

Sand matrix is the binder that holds the sculptures together. Make one batch (enough for the whole class) 1–6 days before you conduct this part. And make a sand sculpture yourself to get the hang of it before doing this part with students.

a. Stir constantly while adding 1 box (3.5 cups) of cornstarch gradually to 3.5 cups of cold water in a saucepan.

b. Heat this mixture over medium heat, *continuing to stir.* Keep stirring until about three-quarters of the matrix has thickened to the consistency of soft mashed potatoes. The rest will be soupy. The mixture will take 5–10 minutes to thicken while on the stove.

c. Remove the mixture from the heat, and stir it until it is consistent and cool. It may get as thick as pudding. Store it in a covered container or jar, or a zip bag.

If the mixture gets too thick to pour, stir in a little water to thin it. Refrigerate the matrix; it will keep up to a week.

4. Prepare sand in basins

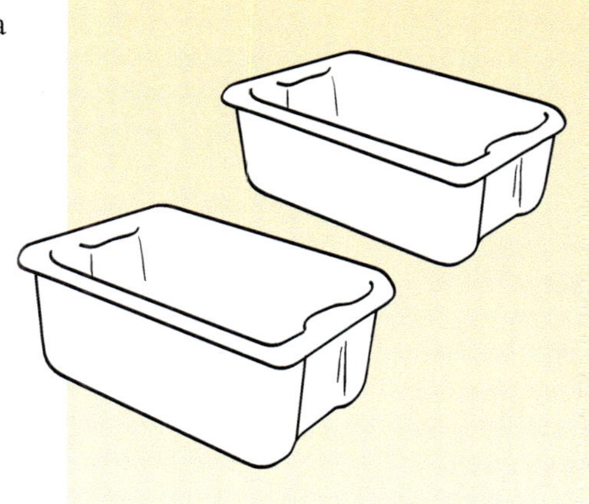

You will need about 2 kilograms (kg) of clean sand. Put it in two basins for easy scooping at activity time. Place several vials in each basin. Each student will scoop one full vial of sand.

5. Plan for drying

Each student will create his or her sculpture on a paper plate. Plan a location for drying. A sunny window will hasten the drying process, which will take 2–3 days.

6. Prepare for cleanup

Have a broom and dustpan handy to clean up spills. The sand-matrix container can be washed in the sink if it contains no sand.

7. Plan assessment: performance assessment

Students determine the number of spoons of sand that make for the best mixture with a given amount of matrix. They verbally make a claim and provide evidence for the amount of sand they used to result in the best mixture for making a sand sculpture.

GUIDING *the Investigation*

Part 3: *Sand Sculptures*

1. Review sand

Call students to the rug. Ask students to share what they have learned about the properties of sand. Write their ideas on chart paper. If needed, ask,

➤ *What is sand made of?* [Small pieces of rock.]

➤ *Is sand smaller or larger than gravel?* [Smaller.]

➤ *Is sand smaller or larger than silt or clay?* [Larger.]

➤ *How is sand like clay? How is it different?* [It is made of small pieces of rock. It doesn't hold together unless you add water. When it dries, it falls apart.]

2. Focus question: How can we make a sand sculpture?

Show the image of the sand castle in the *FOSS Science Resources* book (page 20). Ask students to think about how the designer solved the problem of keeping the small pieces together.

Brainstorm ideas and write them on the board. Students may have experience working with wet sand to build **sculptures** at the beach. Ask students if sand and water sculptures keep their shape over time.

Ask students what they might do to make the sand sculptures permanent. Listen to their engineering ideas.

Ask the focus question and project or write it on the board.

➤ *How can we make a sand sculpture?*

3. Introduce sand matrix

Tell students that a **matrix** is a material that holds particles or materials together. When you add a matrix to sand, the matrix holds the tiny pieces of sand together like glue. The sand and matrix mixture also holds its shape and **hardens** when it dries. Show students the matrix and tell them that you made it from cornstarch and water.

4. Describe the engineering challenge

Explain that each student will receive a paper plate with one measure of matrix on it. Each pair of students will receive a plastic cup half-filled with sand and a metal spoon to share. Each student will need to determine how much sand to add to the matrix to make a good mixture for sculpting.

FOCUS QUESTION

How can we make a sand sculpture?

Materials for Steps 3–4

- *Sand matrix*
- *Paper plates*
- *Basins of sand*
- *Vials*
- *Metal spoon*

TEACHING NOTE

The recipe is not an exact science. Emphasize the need to mix the two materials well. If any student's sand seems a little crumbly, add a little more matrix. If the mixture looks very white, add sand.

SCIENCE AND ENGINEERING PRACTICES

Planning and carrying out investigations

Analyzing and interpreting data

Constructing explanations

DISCIPLINARY CORE IDEAS

PS1.A: Structure and properties of matter

ETS1.C: Optimizing the design solution

CROSSCUTTING CONCEPTS

Scale, proportion, and quantity

Tell students,

Start with 3 level metal spoons of sand and see what the mixture is like. If the mixture is too runny, add one more spoon. Each student can add up to 6 spoons (but not more because we will run out of sand).

5. Demonstrate how to get started

Demonstrate how to measure a level metal spoon of sand. Model how to mix the sand and matrix completely by kneading it. Explain that they can add more measured spoons of sand as needed, up to 6 spoons. Remind students to use their hands to mix the sand and the matrix thoroughly.

6. Distribute paper plates with matrix

Have students move to their tables. Distribute a paper plate to each student. Have them write their names on the outside edge of the plates. Add one metal spoonful of matrix on each plate. Have the Getters get the cups of sand and spoon for each pair of students.

7. Mold sand mixture

Have students begin engineering their mixture and keeping track of the number of spoons of sand they use. When they have a good mixture, they can mold simple shapes from the mixture. They should keep the mixture on the paper plates. When they complete their sculptures, make sure their names are on their paper plates.

8. Assess progress: performance assessment

Conduct 30-second interviews with students as they design their sculptures. Ask questions such as,

- *How many spoons of sand did you use? Would you use the same amount if you did this again? Why or why not?*
- *What would happen if you added [twice] as much sand?*

What to Look For

- *Students describe how many spoons of sand they used, and how they might be able to improve their mixture next time. (Planning and carrying our investigations; analyzing and interpreting data; ETS1.C: Optimizing the design solution.)*
- *Students explain that you have to have a certain amount of matrix to hold the sand together well. Too much sand, and it will crumble. (Constructing explanations; PS1.A: Structure and properties of matter; scale, proportion, and quantity.)*

9. Set sculptures to dry

The sand sculptures will stay on the plates to dry. Have students carefully move their plates to the drying location and then wash their hands. It will take at least 2 days for the sculptures to dry. Any pieces that break off can be reattached with white glue.

10. Review vocabulary

Review key vocabulary added to the word wall during this part.

Have students pair up and take turns thinking of a personal experience they can describe using each of these words. For example, "When I leave my glue stick open it hardens." "In kindergarten we used flour and water to make a matrix for paper mâché paste." "My grandparents have a lion sculpture."

11. Answer the focus question

Have students transcribe the focus question into their notebooks and answer it with words and drawings. They might draw and describe the sculpture they made with sand.

➤ *How can we make a sand sculpture?*

If you have a piece of sandstone available (often called flagstone at garden supply or landscaping stores), have students compare it to their sand sculptures. Can they see the sand grains and the matrix in the sculpture and in the sandstone?

WRAP-UP/WARM-UP

12. Share sand sculptures

Conclude Part 3 or start Part 4 by having students share sand sculptures and notebook entries. Read the focus question together.

➤ *What can we make a sand sculpture?*

Ask students to pair up to

- share their answers to the focus question;
- describe and share their sculptures.

Have students think about and share with their partners what they would do differently next time. To encourage dialogue, provide prompts and questions for students to ask each other, such as,

➤ *What was hard about making it?*

➤ *What would you do differently next time? Why?*

Have students review the reading "Making Things with Rocks" in the *FOSS Science Resources* book. Focus on pages 33–34. Help them to notice the connection between the size of the rocks used and their purpose (larger size for rougher street surface; smaller size for smoother sidewalks).

➤ *How are asphalt and concrete the same and different?*

➤ *If you wanted to make a park for roller skating, what earth materials would you use?*

harden
matrix
sculpture

ELA CONNECTION

This suggested strategy addresses the Common Core State Standards for ELA.

L 5: Demonstrate understanding of word relationships and nuances in word meanings.

Materials for Step 11

- *Piece of sandstone or flagstone (optional)*

ELA CONNECTION

This suggested strategy addresses the Common Core State Standards for ELA.

SL 1: Participate in collaborative conversations.

SCIENCE AND ENGINEERING PRACTICES

Analyzing and interpreting data

CROSSCUTTING CONCEPTS

Cause and effect

No. 14—Notebook Master

MATERIALS *for*

Part 4: *Clay Beads*

For each student

- 1 Ball of potter's clay, about 3 cm (1") in diameter (See Step 4 of Getting Ready.)
- 1 Piece of scratch paper ★
- 1 Piece of yarn, about 60 cm (24") ★
- 1 Piece of straw
- ❑ 1 Notebook sheet 14, *Uses of Earth Materials*

For the class

- 16 Plastic cups
- • Paint brushes ★
- • Clear acrylic spray (optional) ★
- • Felt-tipped pens ★
- • Poster paints ★
- • Tempera paints ★
- 1 Large knife (See Step 4 of Getting Ready.) ★
- 1 Scissors ★
- • Paper towels ★
- • Newspaper ★
- 1 Plastic tablecloth or place mats (optional) ★
- • Water ★
- • Computers with Internet access ★

For embedded assessment

- ❑ • *Embedded Assessment Notes*

★ Supplied by the teacher. ❑ Use the duplication master to make copies.

GETTING READY *for*

Part 4: *Clay Beads*

1. Schedule the investigation

This part will take two active investigation sessions. In the first session, students create clay beads. Students finish their beads 1–2 days later, after the beads have dried.

2. Preview Part 4

Students use clay to make beads or something decorative, which they paint and keep as a memento of their investigation of clay. The focus question is **What makes clay useful in making objects like beads?**

3. Plan for student workspace

There is enough clay for each student to create his or her own bead. If you are concerned about clay on the tables, designate one clay-working site in your classroom where students take turns making their beads. You can cover the table with plastic or use place mats for students to work on.

4. Prepare clay and straws

Each student will need one chunk of potter's clay that will make a ball about 3 cm (1") in diameter. (You will need approximately half a liter of clay.) A large knife is a good tool for cutting clay. You can purchase extra clay at art or school supply stores. Make sure it is real earth clay—not modeling clay or dough clay. Use plastic cups to distribute it.

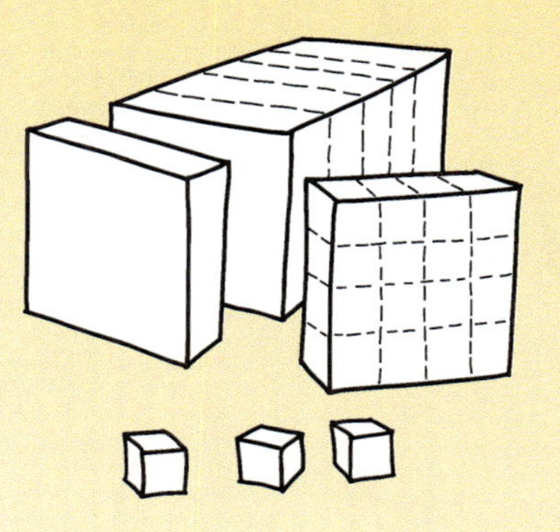

Cut the plastic straws into halves or thirds.

5. Choose a drying area

You will need an area for drying the beads. A sunny windowsill will hasten drying. If a kiln is available, consider firing the beads.

6. Set up a painting area

Plan to set up a painting area in your classroom. Cover the area with newspaper for easy cleanup. To keep the paint from chipping off and to provide a shiny surface, you may want to coat the beads with clear acrylic after the paint has dried. Check with your local arts or craft store for suggestions on what to use.

7. Plan for online activity

Preview the online activity "Find Earth Materials." In this activity, students search for products and objects made of earth materials in two outdoor scenes, one in a park and one along a sidewalk. At any time, students can use the reference to see and read about each rock size: clay, silt, sand, gravel, pebble, cobble, and boulder.

The link to this online activity for teachers is in the Resources by Investigation on FOSSweb.

8. Plan assessment: notebook entry

In Step 8, students write an answer to the focus question. Check students' understanding that clay is good for making beads because the particles are very small and stick together well. When clay dries, it hardens into a very durable object.

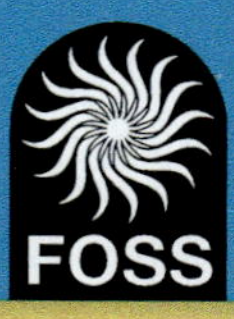

GUIDING *the Investigation*

Part 4: *Clay Beads*

1. Discuss clay properties

Call students to the rug. Ask students, what are the properties of clay? These two important ideas should come forward: (1) it sticks together when it is molded, and (2) it hardens when it dries.

2. Focus question: What makes clay useful in making objects like beads?

Ask the focus question and project or write it on the board.

➤ *What makes clay useful in making objects like beads?*

Brainstorm ideas. Explain that students will use the properties of clay to make a bead decoration.

3. Discuss some techniques

Introduce these techniques to students.

- *If the clay starts to dry out and crack, dip your fingers in water and smooth out the cracks.*
- *Poke a short piece of straw through the clay. Leave it there. Remove it only after the clay dries.*
- *If you like, use a pencil or straw to carve shapes and lines on the clay.*

4. Mold clay

Give each student one chunk of clay and a straw. Have cups of water available. Remind students to work quickly so that the clay doesn't dry out.

Some students may want to make something other than a bead or jewelry. That's OK if you have the appropriate materials for their creations.

5. Store beads to dry

Give each student a piece of scratch paper. Have students write their names on the paper for a label. Have them take their beads and labels to the drying location.

6. Clean up

Have students use the sponges and paper towels to clean up the tables. Students should wash their hands with soap and warm water after this activity.

BREAKPOINT

FOCUS QUESTION

What makes clay useful in making objects like beads?

Materials for Step 4

- *Clay balls*
- *Cups of water*
- *Straw pieces*

Materials for Steps 5–6

- *Scratch paper*
- *Sponges*
- *Paper towels*

Materials for Step 7

- *Painting supplies*
- *Clear acrylic (optional)*
- *Yarn*

TEACHING NOTE

Refer to the Sense-Making Discussions for Three-Dimensional Learning chapter in Teacher Resources on FOSSweb for more information about how to facilitate this with young students.

SCIENCE AND ENGINEERING PRACTICES

Constructing explanations

TEACHING NOTE

Students should circle the bricks, fireplace, and road. Students should write clay soil under bricks; rocks under fireplace; and pebbles and gravel under the road.

7. Finish the project

The clay should dry hard in 1–2 days, depending on the room's humidity. When the clay is hard, have students take turns painting their creations at the painting area. Let the paint dry overnight. Coat the beads with clear acrylic, if you like. Students can thread a piece of yarn through the finished project to make a necklace.

8. Have a sense-making discussion

Use these questions to facilitate a summary conversation.

➤ *What are the properties of clay that make it easy to shape into a bead?* [Small pieces. Sticks together very well.]

➤ *You worked with sand to make sculptures and you worked with clay to make beads. What was different about how you worked with those materials?* [Sand needed a matrix to hold it together. Clay did not. The beads are smoother than the sand sculptures.]

9. Answer the focus question

Have students transcribe the focus question into their notebooks and answer it with words and drawings. They might draw and describe the beads they made with clay.

➤ *What makes clay useful in making objects like beads?*

10. Assess progress: notebook entry

Collect students' notebooks at the end of the session. Check students' explanations about the benefits of using clay to make beads.

What to Look For

- *Students claim that clay is good for beads because the size is so small it sticks together well.*
- *Students indicate that when clay dries, a very durable object is made.*

11. Distribute notebook sheet

Distribute a copy of notebook sheet 14, *Uses of Earth Materials*, to each student. Tell them to circle the objects that they think are made from earth materials. Under each object is a space for them to write what the object is made of, such as clay, sand, or pebbles.

12. View online activity: "Find Earth Materials"

In small groups or as individuals, have students engage with the online activity focusing on how rocks are used in our everyday life.

WRAP-UP/WARM-UP

13. Share clay sculptures

Conclude Part 4 or start Part 5 by sharing clay beads. Read the focus question together.

➤ *What makes clay the best earth material for making beads?*

Ask students to work with a partner to share their response to the focus question and discuss these questions.

- compare the clay beads to the sand sculptures;
- explain if they could (or could not) make a bead out of sand.

ELA CONNECTION

This suggested strategy addresses the Common Core State Standards for ELA.

SL 1: Participate in collaborative conversations.

CENTER INSTRUCTIONS—MAKING BRICKS

Materials

Bucket of clay soil	Dry grass or straw	Scratch paper
Plastic cups	Aluminum pans	Pencils
Basins	Container of water	Paper towels

Set Up the Center

Find a place outdoors to set up the center. Gather the materials and take them outside. A pair of students will make one brick. A group of four students will work around one basin.

Guide the Investigation

1. **Introduce the clay soil.** Call students around the bucket of clay soil. Have them rub some of it between wet fingers. Tell them this soil has a lot of clay.
2. **Prepare the mud.** Pairs of students work together to make one brick. Have students roll up their sleeves and follow this procedure.
   a. Each student measures about two cups of clay soil into a basin.
   b. Add water while students mix with their hands. They should break up the soil lumps and remove rocks. Proper brick mud should be sticky and too thick to pour.
   c. Each student adds one small handful of grass or weeds, working it in thoroughly.
3. **Mold the bricks.** When the mud is ready, distribute aluminum pans to pairs of students and direct them to mold their bricks.
   a. Have students use their hands to put mud into the loaf pans. The pans should be about two-thirds full.
   b. Students should press the mud uniformly into the pan and pat the surface flat and smooth.
4. **Dry the bricks.** Tell students that the bricks have to "bake" several days in a warm, dry location. Have them carefully move the bricks to the drying location. They should place their pans on scratch paper on which they have written their names.
5. **Clean up.** Have students wash out the basins. Fill a clean basin with water and have them wash their hands. The last group should dispose of the leftover clay soil in a garden area, and rinse the bucket and basin.

FOSS Next Generation

FOSS Pebbles, Sand, and Silt Module
Investigation 3: Using Rocks
No. 13—Teacher Master

No. 13—Teacher Master

MATERIALS *for*

Part 5: *Making Bricks*

For each pair of students

- 1 Aluminum mini loaf pan
- • Clay soil, about 1/2 L (See Step 4 of Getting Ready.) ★
- 1 Handful of dry grass, straw, or weed clippings ★
- 1 Half sheet of scratch paper ★
- 2 *FOSS Science Resources: Pebbles, Sand, and Silt*
  - "What Are Natural Resources?"

For the class

- 8 Basins
- 2 Plastic cups
- 1 Brick (optional) ★
- 1 Trowel or shovel ★
- • Containers for water ★
- • Water ★
- • Paper towels or old hand towel ★
- 1 Bucket ★
- ❑ 1 Teacher master 13, *Center Instructions—Making Bricks*
- 1 Big book, *FOSS Science Resources: Pebbles, Sand, and Silt*

For benchmark assessment

- ❑ • *Investigation 3 I-Check*
- • *Assessment Record*

★ Supplied by the teacher.

❑ Use the duplication master to make copies.

GETTING READY *for*

Part 5: *Making Bricks*

1. Schedule the investigation

This part will take two active investigation sessions, one session for the reading, and one session for the I-Check. This first active investigation session will take 15 minutes for discussion and 15–20 minutes per group to make the bricks. The bricks need to dry 2 or more days before they can be dumped out of the pans. A second active investigation session, conducted when the bricks are completely dry, will take 25–30 minutes. Consider the weather, as well. Warm, dry weather is best for drying the bricks.

2. Preview Part 5

Students make adobe clay bricks with a mixture of clay soil, dry grass or weeds, and water. After the bricks dry, they can be used to build a class wall. The focus question is **How are bricks made?**

3. Enlist additional adults

It is very helpful to have an additional adult or two to help with brick making. Use the teacher master *Center Instructions—Making Bricks* to help guide your assistant at the center.

4. Dig up clay soil

Locate a source of clay soil in your area and dig up a bucket full. Clay soil sticks together when made into a mud ball, even when flattened.

If you need help finding clay soil, contact your regional geological survey. You can also purchase and add powdered clay to backyard soil to create a suitable medium.

5. Collect grass and weeds

You need dry grass, weeds, or straw to add to the brick mixture. Grass and straw work best.

6. Acquire a brick

If possible, bring a commercially produced brick to class to use in Step 1 of Guiding the Investigation.

7. Check aluminum mini loaf pans

Mini loaf pans are used to form the bricks. They can be reused, so check the condition of the ones in the kit. Provide one pan for each pair of students.

8. Select outdoor site

Brick making is best done outdoors, at a center with a group of four students to a basin. You can limit the number of groups making bricks at one time or have all the groups do it together. Find a spot in your schoolyard where it's OK to make mud, water is nearby, and it is easy to clean up. See the Taking FOSS Outdoors chapter for tips on bringing water outdoors.

Bricks dry faster in a sunny, dry location. If the air-conditioning is on in your classroom and it is hotter and drier outdoors, consider drying your bricks outdoors in a location away from student traffic.

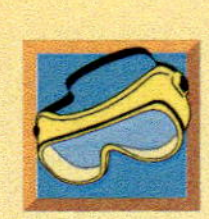

9. Check the site

Tour the outdoor brick-making site on the morning of the activity. Check the area for unsightly and distracting items.

10. Provide mortar for building walls (optional)

You can make some mortar to stick the finished bricks together. Follow the recipe for the sand matrix described in Part 3.

11. Plan for cleanup

Fill one basin with water for students to use to rinse the clay off their hands. Provide paper towels or an old hand towel for drying hands and tools.

Some teachers prefer to use an assembly line to clean up. Place a basin of water for students to clean the mud off their hands, then a bucket for rinse, paper towels or an old hand towel, and a trash can.

12. Plan to read *Science Resources*: "What Are Natural Resources?"

Plan to read "What Are Natural Resources?" during a reading period before the benchmark assessment session for this part.

13. Plan assessment: I-Check

Plan to give *Investigation 3 I-Check* at the end of the investigation. Read the items aloud to the whole class, and have students answer independently. Review students' responses using the What to Look For information in the Assessment chapter. Use assessment master 3, *Assessment Record,* to record students' responses.

GUIDING *the Investigation*

Part 5: *Making Bricks*

1. Introduce bricks

Call students to the rug. Tell them that people use earth materials in many ways. One use of clay soil is to make bricks. Show them the manufactured brick if you have one. Open *FOSS Science Resources* book to page 36 and have the students study the bricks in the wall.

2. Focus question: How are bricks made?

Ask the focus question and project and write it on the board.

➤ *How are bricks made?*

Brainstorm ideas. Lead the class through a process of asking questions about what materials they need (clay soil). They can look at images of adobe bricks or an actual brick to get ideas.

Have students discuss the process and suggest where they might find the best clay soil and dead grass or straw for the materials. Ask them to describe how they would mix the materials with water and what kind of mold they might use to form the bricks.

3. Go outdoors

Identify eight to ten brick makers or have the whole class go outdoors together. Together, carry the supplies to the brick-making location.

4. Introduce the clay soil

Show students the clay soil. Have them rub some of it between wet fingers. Tell them this soil has a lot of clay. Ask,

➤ *Why do you think we need material with lots of clay for making bricks?* [Bricks need to be hard when they dry.]

5. Prepare the mud

Have students roll up their sleeves and follow this procedure.

a. Each student in the group of four students measures about two cupfuls of clay soil into a basin.

b. Add water while students mix with their hands. They should break up the soil lumps and remove rocks. Proper brick mud should be sticky and too thick to pour.

c. Each student adds one small handful of grass, straw or weeds, working it in thoroughly.

FOCUS QUESTION

How are bricks made?

Materials for Step 1

- *Brick (optional)*

Materials for Step 3

- *Bucket of clay*
- *Basins*
- *Water*
- *Dry grass or weeds*
- *Plastic cups*
- *Aluminum pans*

SCIENCE AND ENGINEERING PRACTICES

Constructing explanations

Materials for Steps 7–8
- *Water*
- *Paper towels*
- *Scratch paper*

6. Mold the bricks

When the mud is ready, distribute an aluminum pan to each pair of students and direct them to mold their bricks.

a. Have students use their hands to put mud into the loaf pans. The pans should be about two-thirds full.

b. Students should press the mud uniformly into the pans and pat the surface flat and smooth.

7. Dry the bricks

Tell students that the bricks have to "bake" several days in a warm, dry location. Have each pair of students carefully move their brick to the drying location. Each pair should place their pan on a half sheet of paper on which they have written their names.

8. Clean up and return to class

Have students help wash out the basins to cleanup. Fill a clean basin with water so they can wash their hands. Don't pour any soil mixture down a drain; recycle it outdoors in a garden.

BREAKPOINT

9. Remove bricks from mold

The bricks should be dry enough after a day or two to remove them from the pans and to place them on scratch paper. Test one of the bricks before you let students remove them.

Reshape and wash the pans, if necessary, so they are ready for the next set of brick makers.

Materials for Step 10
- *Sand matrix (optional)*

CROSSCUTTING CONCEPTS

Energy and matter

10. Use the bricks

It will probably take at least 1 week for the bricks to dry thoroughly, depending on the warmth and humidity of the drying location. When the bricks are hard, have students write their names on the flat side with a pencil. Have them stack their bricks together in a wall or some other construction. Sand matrix can be used as mortar to stick the bricks together. This is a good time to reinforce that rock particles can be put together into larger pieces to form bricks (energy and matter crosscutting concept).

11. Answer the focus question

Have students transcribe the focus question into their notebooks and answer it with words and drawings. They might draw and describe the bricks they made with clay soil.

➤ *How are bricks made?*

READING *in Science Resources*

12. Read "What Are Natural Resources?"

Point out the title of the article and ask students what they think the article is going to be about. Do a think-pair-share about what they think **natural resources** are. Encourage students to think about what each word means individually and what they might mean together.

Have students pair up and take turns reading one page to the other. The student listening answers the questions.

Next, read the article aloud and pause for students to indicate which structures are natural and which are made by people. Ask students to explain their reasoning.

Have students look over the pictures again with a partner and talk about a rock wall, walking paths, and rock gardens.

- *Which ones are natural?*
- *Which ones are made by people?*
- *How do you think making these walls or pathways affects humans or other living things?*

Ask students to talk with a partner about a rock wall, walking path, or rock garden at their school or in their community.

13. Have a sense-making discussion

Ask students if they have a better idea about what natural resources are. Model how to use a Frayer Model on chart paper to deepen their understanding of this concept.

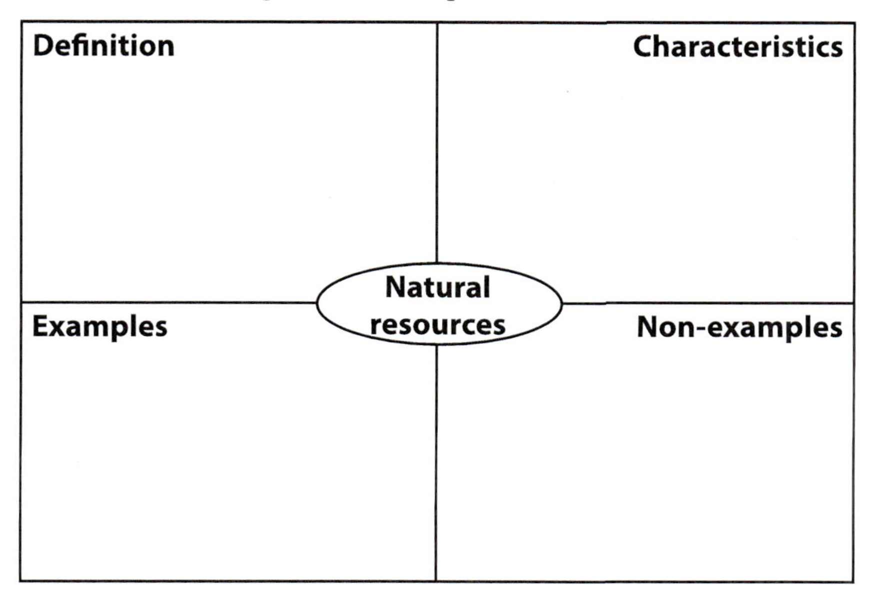

What Are Natural Resources?

Rocks are **natural resources**. Rock walls can be formed by nature. Rock walls can be made by people, too.

Look at the rock walls. Which ones are natural? Which ones are made by people?

38

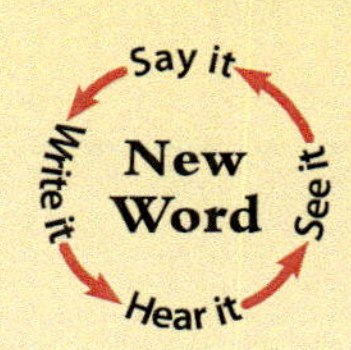

SCIENCE AND ENGINEERING PRACTICES

Obtaining, evaluating, and communicating information

ELA CONNECTION

These suggested strategies address the Common Core State Standards for ELA.

RI 1: Ask and answer questions to demonstrate understanding.

RI 3: Describe the connection between scientific ideas or concepts.

RI 5: Know and use text features.

L 4: Determine of clarify the meaning of unknown or multiple-meaning words and phrases.

Start with a definition. Students can look up the word in the glossary to confirm their understanding. Together craft a definition such as, "things we get and use from Earth."

Next, have students come up with examples. [Soil, rocks, water, air.] Then let students think about and share some non-examples. [Plastic, fabric, rubber.]

Finally, note the characteristics of the natural resources as used in the context of this module. [Come from non-living things, not made by humans.]

14. Revisit the guiding question

To revisit the guiding question about how earth materials are used by humans to make things, you can take a second trip outdoors to the schoolyard or have students review the *FOSS Science Resources* book, pages 32–40, and then ask a few questions.

- ➤ *Where are small pieces of earth material mixed together to make larger objects?*
- ➤ *Where are earth materials used to make smooth objects or surfaces?*
- ➤ *What are the ways that earth materials are put together to make useful things at our school?*

SCIENCE AND ENGINEERING PRACTICES

Constructing explanations

CROSSCUTTING CONCEPTS

Energy and matter

BREAKPOINT

15. Assess progress: I-Check

When students have completed the investigation, give them *Investigation 3 I-Check.*

Review student responses. Use the What to Look For information in the Assessment chapter for guidance. Note concepts that you might want to revisit with students, using the next-step suggestions.

The students' experiences in this investigation contribute to their understanding that matter can be described by its observable properties, that different properties are suited to different purposes, and that a great variety of objects can be built up from a small set of pieces; and that the engineering design process involves defining problems, developing possible solutions, and comparing solutions to optimize the design solution.

DISCIPLINARY CORE IDEAS

PS1.A: Structure and properties of matter

ETS1.A: Defining and delimiting engineering problems

ETS1.B: Developing possible solutions

ETS1.C: Optimizing the design solution

INTERDISCIPLINARY EXTENSIONS

Language Extensions

- **Find out about pottery**
  Invite a potter to class or visit a potter's studio to find out how clay is used to produce works of art and kitchenware. Have students prepare interview questions for the guest speaker. This might be an opportunity to have students' clay work glazed and fired.
- **Make tracks and molds**
  Supply some extra clay for students to make tracks or molds. The clay should be fairly soft for making hand or foot impressions. Students can also push shells and other hard objects in the clay to leave impressions or molds. Have students write about the object they used to make the impression.
- **Look for rocks everywhere**
  Have students continue their search for rocks and their uses. Ask them to search for rocks in use at their homes and in the outdoors. Make a class list of the examples students discover. Add to the list each day as students offer more examples. Adding to the rock list can be incorporated into the morning routine for a week or so.

  Discuss how the size of a rock affects its use. Buildings are built with large stones, not with clay. Sand, not boulders, is used in a play area. As a group or in pairs, have students complete the sentence frames below. Add or change the sentence frame to fit your class's needs.

  If I wanted . . .

  to build a tower, I would use _____.

  to make beads, I would use _____.

  to make a walking path, I would use _____.

  to make a pot, I would use _____.

  to make a wall, I would use _____.

  to make a play area, I would use _____.

TEACHING NOTE

Refer to the teacher resources on FOSSweb for a list of appropriate trade books that relate to this module.

Name ________________ Date __________

MATH EXTENSION A

Investigation 3: Using Rocks

You and five friends want to make clay bricks at home.

1. Each person needs 2 cups of clay. Altogether, how many cups of clay will you need?

2. Each person needs 1 handful of straw. How much straw will you need for everyone?

3. Every two people share 1 cup of water. How many cups of water will you need in all?

No. 14—Teacher Master

Math Extensions

- ### Math Problem A

Six children are going to make bricks with the same ingredients your students used. These children want to be sure they have enough of each ingredient for everyone to make one brick. Students determine the number of cups of clay and water as well as handfuls of straw needed.

Notes on the problem. Have students work with partners and present their strategies to solve the problem. It is likely they will draw pictures, and some may use tally marks. In addition, it may be a good opportunity to introduce a T-table where they can keep track of the amounts needed by the number of children as follows.

Children	Cups of clay
1	2
2	4
3	6
4	8
5	10
6	12

The six children need 12 cups of clay (2:1 correspondence).

Children	Handfuls of straw
1	1
2	2
3	3
4	4
5	5
6	6

The six children need 6 handfuls of straw (1:1 correspondence).

Children	Cups of water
1	1/2
2	1/2 + 1/2 = 1
3	1 + 1/2 = 1 1/2
4	1 1/2 + 1/2 = 2
5	2 + 1/2 = 2 1/2
6	2 1/2 + 1/2 = 3

The six children need 3 cups of water (1/2:1 correspondence).

- **Math Problem B**
  In Problem 1, two people are building a brick wall and they run out of bricks. Each row is made of six bricks. The wall is four bricks tall. How many more bricks will they need for each row? How many more bricks will they need for the entire wall?

  Notes on the problems. Be sure that students understand that the drawing shows some bricks in place and some bricks missing. The outlines represent missing bricks. If they need help getting started, point out that the bottom row of six bricks is complete.

  Problem 2 is more difficult and students will have to draw their own brick wall. This new wall will require 35 bricks total. Since the people have 10 bricks, they will need 25 more.

- **Make clay dice**
  Have students fashion dice out of a small piece of clay. Dry the dice and use them in math games.

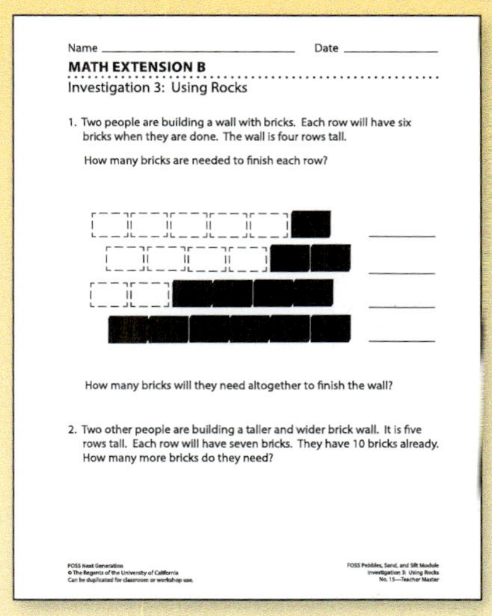

Name ______________ Date ______________

MATH EXTENSION B

Investigation 3: Using Rocks

1. Two people are building a wall with bricks. Each row will have six bricks when they are done. The wall is four rows tall.

   How many bricks are needed to finish each row?

   How many bricks will they need altogether to finish the wall?

2. Two other people are building a taller and wider brick wall. It is five rows tall. Each row will have seven bricks. They have 10 bricks already. How many more bricks do they need?

No. 15—Teacher Master

Engineering Extension

- **Look at construction materials**
  Find a local building supply store or aggregate supplier. Take students on a field trip to find out what kinds of construction materials are made from earth materials, how they are packaged, and how they are used.

TEACHING NOTE

Encourage students to use the Science and Engineering Careers Database on FOSSweb.

Art Extensions

- **Make sand paintings**
  Have students create sand paintings. Fill a basin with sand you have acquired. Have students follow this procedure.

  a. *Quickly draw a picture on a piece of black construction paper with a glue stick or thin lines of white glue.*

  b. *Turn the paper face down 1 minute in the sand and apply a little pressure with the palm of your hand.*

  c. *Pick up the picture and shake any excess sand back into the basin.*

  Sand will be attached to the glue. Allow the "paintings" to dry. Students can add pebbles and gravel to complete the art. Have students find out how some Native American cultures create and use sand paintings.

- **Layer sand, gravel, and pebbles**
  Have students place layers of sand, gravel, and pebbles in a jar to create an interesting effect. You can add color to the sand by mixing in powdered tempera paint or crushed colored chalk.
- **Make sandpaper prints**
  Have students draw with crayons on extra pieces of sandpaper. Have an adult help students iron the drawings onto white paper. Turn the crayon side of the sandpaper toward the white paper and press a hot iron on the back to transfer the design to the white paper.

Science Extensions

- **Research animals and earth materials**
  Have students find out about animals that use earth materials in building their homes or that live in the ground. Some suggestions are beavers, termites, wasps, birds, gophers, moles, and worms.

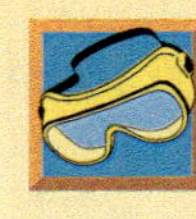

- **Sand wood**
  If you would like students to have more experience of sanding wood, acquire some wood and some extra sandpaper. Have them do their sanding over newspaper or paper plates for easy cleanup. Or take students outdoors to find a stick to sand.

Environmental Literacy Extension

- **Make paving stone habitats**
  Ask the grounds manager at the school to help your class design paving stones that can be placed at the edge of a schoolyard garden or to line a pathway. Over time, these small tiles will become natural habitats for worms, isopods, snails, slugs, and other small organisms. A small bag of cement can be used to make the tiles and students will gather and add stones, shells, or glass beads to decorate the tiles. (This is an investigations in the **FOSS Soils, Rocks, and Landforms Module** for grade 4, so the upper-grade students could serve as mentors to the second-grade class.)

Home/School Connection

Students become field-trip leaders as they take their families on a search for rocks in use. A family member helps record all the places where they see rocks in use.

Print or make copies of teacher master 16, *Home/School Connection* for Investigation 3, and send it home with students after Part 1.

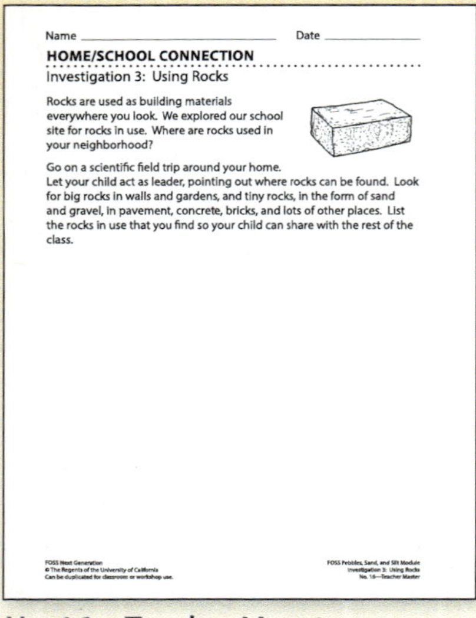

Name ____________ Date ____________

HOME/SCHOOL CONNECTION

Investigation 3: Using Rocks

Rocks are used as building materials everywhere you look. We explored our school site for rocks in use. Where are rocks used in your neighborhood?

Go on a scientific field trip around your home.
Let your child act as leader, pointing out where rocks can be found. Look for big rocks in walls and gardens, and tiny rocks, in the form of sand and gravel, in pavement, concrete, bricks, and lots of other places. List the rocks in use that you find so your child can share with the rest of the class.

FOSS Next Generation
© The Regents of the University of California
Can be duplicated for classroom or workshop use.

FOSS Pebbles, Sand, and Silt Module
Investigation 3: Using Rocks
No. 16—Teacher Master

No. 16—Teacher Master

Investigation 4: Soil and Water

INVESTIGATION 4 — *Soil and Water*

Guiding question for phenomenon:
How can we apply what we know about the ways land and water interact?

PURPOSE

Students first investigate a common phenomenon on the surface of Earth—soil. After systematic analysis, students find that soil is composed of rock particles of various sizes and organic material, humus. Students investigate interactions of soil with another earth material—water—and focus on ways to reduce erosion.

Content

- Soils can be described by their properties. Soil is made partly from weathered rock and partly from organic material. Soils vary from place to place.
- Natural sources of water include streams, rivers, ponds, lakes, marshes, and the ocean. Sources of water can be fresh or salt water. Water can be a solid, liquid, or gas.
- Wind and water can cause erosion of rock and soil.

Practices

- Find, collect, record, and compare samples of soil outside the classroom. Use water to separate soil parts.
- Compare engineering designs to prevent erosion.
- Compare models of land and water to identify common features and differences (photographs, drawings, maps).

Science and Engineering Practices

- Asking questions and defining problems
- Developing and using models
- Planning and carrying out investigations
- Analyzing and interpreting data
- Constructing explanations and designing solutions
- Engaging in argument from evidence
- Obtaining, evaluating, and communicating information

Disciplinary Core Ideas

ESS1: What is the universe, and what is Earth's place in it?
ESS1.C: The history of planet Earth
ESS2: How and why is Earth constantly changing?
ESS2.A: Earth materials and systems
ESS2.B: Plate tectonics and large-scale system interactions
ESS2.C: The roles of water in Earth's surface processes
ETS1: How do engineers solve problems?
ETS1.A: Defining and delimiting engineering problems
ETS1.B: Developing possible solutions
ETS1.C: Optimizing the design solution

Crosscutting Concepts

- Cause and effect
- Scale, proportion, and quantity
- Stability and change

INVESTIGATION 4 — *Soil and Water*

	Investigation Summary	Time	Focus Question for Phenomenon, Practices
PART 1	**Homemade Soil** Students put together and take apart soils. They are introduced to humus, an important soil ingredient. They mix together homemade soil containing sand, gravel, pebbles, and humus. They shake some of the soil on a paper plate and observe what happens. They use screens to separate the homemade soil. They shake soil and water together in a vial and draw what they observe.	**Active Inv.** 3 Sessions *	**What is soil?** **Practices** Asking questions and defining problems Planning and carrying out investigations Analyzing and interpreting data Constructing explanations Engaging in argument from evidence
PART 2	**Local Soil** Students go on a schoolyard field trip to collect soil samples. They try to find soil in as many places as possible: next to sidewalks, near trees, and in landscaped areas. Students study their schoolyard soil samples. They shake vials with the soil and water, then draw the results. They compare the vials and drawings of their schoolyard samples with the vials and drawings of the homemade soil.	**Active Inv.** 3 Sessions **Reading** 2 Sessions	**How do soils differ?** **Practices** Asking questions and defining problems Planning and carrying out investigations Analyzing and interpreting data Constructing explanations Engaging in argument from evidence Obtaining, evaluating, and communicating information
PART 3	**Natural Sources of Water** Students read about sources of natural water, sort images of water sources, both fresh and salt, and discuss where water is found in their community.	**Reading** 2 Sessions	**Where is water found in our community?** **Practices** Obtaining, evaluating, and communicating information
PART 4	**Land and Water** Students compare a variety of solutions to slow down the effects of wind and water erosion on land. They go out on the schoolyard to look for erosion. They end the module by studying a variety of images, representing different landforms and bodies of water, and identify common features and differences.	**Reading** 2 Sessions **Assessment** 1 Session	**How can soil erosion be reduced?** **Practices** Developing and using models Constructing explanations and designing solutions Obtaining, evaluating, and communicating information

* A class session is 45–50 minutes.

Content Related to DCIs	Writing/Reading	Assessment
• Humus is decayed material from plants and animals. • The ingredients of soil can be observed by mixing soil with water, shaking it, and letting it settle. • Soil is made partly from weathered rock and partly from organic material.	**Science Notebook Entry** *Homemade Soil with Water*	**Embedded Assessment** Performance assessment
• Soils can be described by their properties (particle size, color, texture, ability to support plant growth). • Soil is made partly from weathered rock and partly from organic material. • Soils vary from place to place. • Soils differ in their ability to support plants.	**Science Notebook Entry** *Local Soil with Water* **Science Resources Book** "What Is in Soil?" "Testing Soil" **Video** *All about Soil*	**Embedded Assessment** Science notebook entry
• Earth materials are natural resources. • Natural sources of water include streams, rivers, ponds, lakes, marshes, and oceans. Sources of water can be fresh or salt water. • Water can be a solid, liquid, or gas.	**Science Notebook Entry** *Water in Our Community* *States of Water* **Science Resources Book** "Where Is Water Found?" "States of Water"	**Embedded Assessment** Science notebook entry
• The shapes and kinds of land and water can be represented in photos, drawings, and maps. • Wind and water change the shape of the land. • Engineers design methods to slow erosion by wind and water.	**Science Notebook Entry** Answer the focus question **Science Resources Book** "Erosion" "Ways to Represent Land and Water" **Video** *All about Landforms* (review)	**Embedded Assessment** Science notebook entries **Benchmark Assessment** *Investigation 4 I-Check* **NGSS Performance Expectations addressed in this investigation** 2-ESS1-1 2-ESS2-1; 2-ESS2-2; 2-ESS2-3 K–2 ETS1-1; K–2 ETS1-2; K–2 ETS1-3

BACKGROUND *for the Teacher*

What Is Soil?

What is the phenomenon called **soil**? To the farmer, soil is the layer of earth material in which plants anchor their roots and from which they get the nutrients and water they need to grow. To a geologist, soil is the layer of earth materials at Earth's surface that has been produced by weathering of rocks and sediments and that hasn't moved from its original location. To an engineer, soil is any ground that can be dug up by earth-moving equipment and requires no blasting. And to students, soil is dirt.

In FOSS, *soil* is defined as a mixture of different-sized earth materials, such as gravel, sand, and silt, and organic material called **humus**. Humus is the dark, musty-smelling stuff derived from the **decayed** and decomposed remains of plant and animal life. The proportions of these materials that make up soil differ from one location to another.

How Do Soils Differ?

The ideal composition of a soil to grow most plants would have a recipe like this.

25%	air
25%	water
45%	earth materials (e.g., sand, silt, and clay)
5%	humus

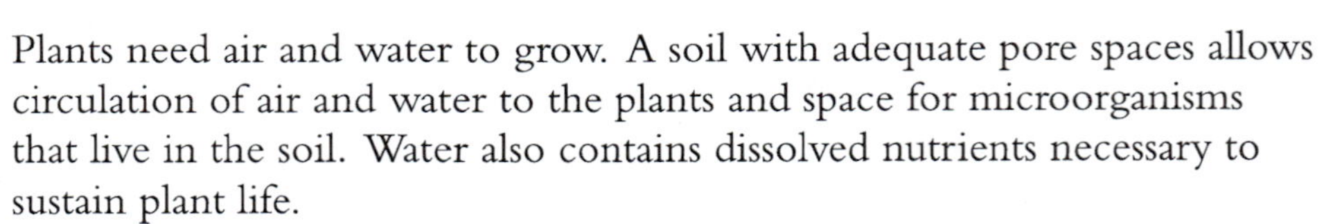

Plants need air and water to grow. A soil with adequate pore spaces allows circulation of air and water to the plants and space for microorganisms that live in the soil. Water also contains dissolved nutrients necessary to sustain plant life.

You'll notice that the amount of humus in the ideal soil is considerably less than the other three components. But humus has an important role in soil. It is a source of nutrients for plant growth, and it increases the soil's ability to absorb and **retain** water. If you were to pour equal amounts of water through two flowerpots, one filled with sand and the other filled with soil, you'd find that more water collects in the saucer under the sand-filled pot. More water is absorbed by the soil.

Geologists classify soils by the amounts and kinds of organic and earth materials that compose them. A soil rich in organic material might be called a bog soil or a histosol. Loam is soil containing a mixture of sand, silt, and clay. Loams can be further described by their major components, for example, sandy loam or silty loam.

All life depends on a dozen or so elements that must ultimately be derived from Earth's crust. Soil has been called the bridge between earth material and life; only after minerals have been broken down and incorporated into the soil can plants process the nutrients and make them available to humans and other animals.

Where Is Water Found in Our Community?

Water. The word brings to mind many images and familiar experiences, from the colorless, tasteless, transparent liquid filling our kitchen sinks to the soothing rhythm of waves rolling up on a beach. Water is, on one hand, the most common, most readily identifiable earth material that we encounter daily. After all, it covers almost three-fourths of Earth's surface. On the other hand, water is a strange, unique material with properties quite unlike any other earth material.

Water is the only substance on Earth that is found naturally in all three states of matter: **solid**, **liquid**, and **gas**. It can change from one state to another, depending on the amount of heat that is added or taken away from it. Liquid water freezes to solid ice at 0°C, and ice melts when it is warmed above this temperature. Water becomes an invisible gas, or vapor, when it reaches 100°C.

Earth is a water-rich planet, but 97% of the planet's water is in a huge **saltwater ocean**, and in smaller seas, bays, and saltwater marshes. That means only 3% of the planet's water is fresh. Unfortunately for freshwater fans, about two-thirds of it is frozen in polar caps and glaciers. That leaves just 1% of the planet's water as liquid **fresh water**.

Liquid fresh water is found in many places. A lot of water is underground (about 95% of the liquid water is underground). Most of the rest is on the planet's solid surface. We see it in **ponds**, **lakes**, **rivers**, and **streams**. We watch it fall to Earth as rain and observe it as fog and dew. We can't see it, but water is in the air as water vapor gas. There is more water in the air than in all the rivers on Earth.

Fresh water is distributed in the atmosphere and on the land unevenly. Consequently, water is absurdly abundant in some parts of the world and scarce to the point of being precious in other places. Water is an essential natural resource for human life and is a renewable resource that needs to be carefully used and conserved. We drink water and use it for countless other personal and community activities and for nourishing our food sources.

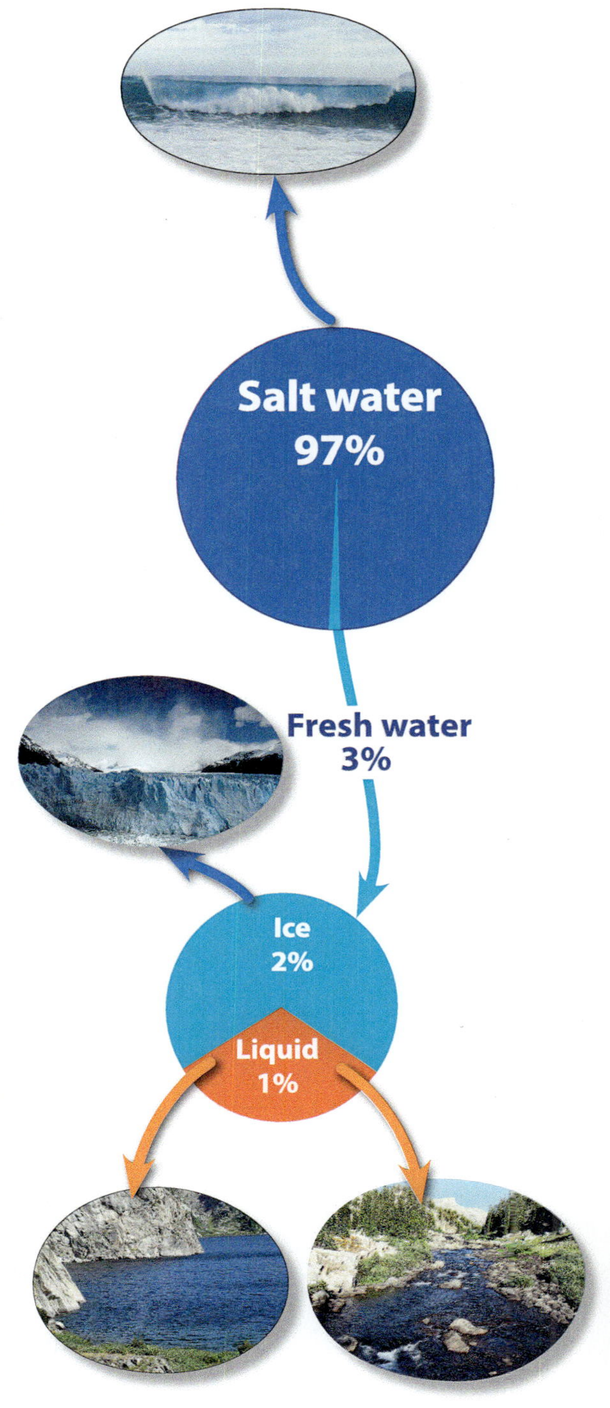

Photo credits: Glacial ice photo: © iStockphoto/Warwick Lister-Kaye
All other photos: © Larry Malone/Lawrence Hall of Science

How Can Soil Erosion Be Reduced?

Erosion is a natural process that has been occurring on Earth throughout its entire history, since the chaotic initial phase of Earth's formation. Weathering forces break rock into progressively smaller pieces (sediments) until they are susceptible to relocation due to the forces exerted by moving water and air. The movement or transport of sediments by wind and water is erosion. Erosion in and of itself is not a bad thing, but when human activities exacerbate the rate or magnitude of erosion, to the point of potential disruption of human activities, erosion can rise to the level of a problem. Erosion issues arise particularly at times of extreme weather. When exceptional storms create huge surge waves that batter coastlines, shoreline erosion can undercut the soil on which humans have built homes, businesses, and other artifacts of human societies. Similarly, intense seasonal rain and snow melt can bring inordinate volumes of water into rivers systems (flood), causing them to claw away at the banks. The challenge is keeping rivers constrained to their natural courses. And any sloped bare land is vulnerable to soil erosion if it is exposed to intense rain, or even moderate rain over an extended period of time.

In places where the potential for damaging erosion has been identified, measures can be taken to reduce or prevent catastrophic erosion. In some instances, erosion prevention can be accomplished by establishing or maintaining a healthy growth of natural shoreline vegetation. Wetlands are now recognized for their value as stabilizers of vulnerable coastlines and riverbanks. Where coastal wetlands have been replaced with development, humans often resort to engineered resistance to counter erosion. Great quantities of large boulders, known as riprap, are often placed along shorelines to resist the powerful assault of surging water. This inelegant form of engineering is often effective at shielding vulnerable coastal soil from erosion in all but the most extreme conditions. Similar coastline and riverbank solutions may involve the deployment of sand bags, tiles, or manufactured interlocking concrete shapes to establish erosion barriers on vulnerable soil facing the ravages of moving water.

Hillsides that have been denuded as a result of construction, poor agricultural practice, or fire, can recover their erosion resistance if they are able to reestablish their natural vegetation. Measures to assist in this process involve engineering runoff barriers across the sloped area to slow the speed of the water flowing down the slope, and applying seed-impregnated net-like blankets over the surface to encourage the plant cover regeneration process.

Hiking trails that are susceptible to erosion are engineered in a number of ways to maintain the integrity of the trail. Flow diverters positioned across a section of trail direct water flowing down the trail onto the surrounding land where it can flow away slowly or soak into the ground harmlessly. In more extreme cases, sections of trail are sometimes replaced with concrete steps over which water can flow without causing erosion. Over less precipitous sections, a trail is occasionally replaced with an elevated boardwalk.

The truth is that most of the steps taken by people to protect against soil erosion are designed to protect against fairly routine assaults by moving water. When truly extreme catastrophic climatic events conspire to attack the soil (hurricanes, 100-year floods,etc.), there is very little humans can do to resist the erosive forces that ensue.

Representing landforms and waterways. The best way to appreciate the full impact of the natural world is to wander around in it and engage it with the full complement of your human senses. But, of course it is not possible to visit every place on Earth for which we might want or need to have knowledge. Over time humans have developed a number of traditional methodologies for representing information about natural structures and systems for which we typically will never have opportunity to experience personally. Early natural historians relied on drawings and paintings to capture critically important aspects of the natural wonders they encountered.

A representation of the relative positions of things of interest in a reference frame is a map. Mapmaking produces a different kind of information about a natural location—not what it looks like, exactly, but an idea of where it is and how it relates to other familiar points of reference. The locations of large features or identifying structures is captured in maps.

In the 19th century the advent of photography gave the intrepid naturalist explorers a new technology for recording representations of natural places. First still monochrome images, then not too long after, moving images, and finally full color still and moving images.

These three classes of representation of the natural world—photographic images, maps, and drawings—when embraced in unison can provide a person with a kind of primitive model of the system being represented. A fully formed scientific model is an explanatory concept, an effort to explain simultaneously the structure, activity, and implication of a complex system. A photographic image, painting, and map of Crater Lake together provide a lot of information about the place, but fall short of a complete scientific model.

TEACHING CHILDREN *about Soil and Water*

Developing Disciplinary Core Ideas (DCI)

NGSS Foundation Box for DCI

ESS1.C: The history of planet Earth
- Some events happen very quickly; others occur very slowly over a time period much longer than one can observe. (2-ESS1-1)

ESS2.A: Earth materials and systems
- Wind and water can change the shape of the land. (2-ESS2-1)

ESS2.B: Plate tectonics and large-scale system interactions
- Maps show where things are located. One can map the shapes and kinds of land and water in any area. (2-ESS2-2)

ESS2.C: The roles of water in Earth's surface processes
- Water is found in the ocean, rivers, lakes, and ponds. Water exists as solid ice and in liquid form. (2-ESS2-3)

ETS1.A: Defining and delimiting engineering problems
- Before beginning to design a solution, it is important to clearly understand the problem. (K-2-ETS1-1)

ETS1.B: Developing possible solutions
- Designs can be conveyed through sketches, drawings, or physical models. These representations are useful in communicating ideas for a problem's solutions to other people. (K-2-ETS1-2)

ETS1.C: Optimizing the design solution
- Because there is always more than one possible solution to a problem, it is useful to compare and test designs. (K-2-ETS1-3)

To most primary students the thin layer of Earth where they dig holes and where plants grow is called dirt. And dirt has not usually been afforded the noble station it deserves. Primary students associate dirt with the unwelcome condition of their clothes after a day of youthful enterprise. Dirt is scorned when it is tracked into the home after play. Dirty things are generally considered bad. This investigation attempts to reverse the image of this essential component of the natural world.

We refer to this misunderstood material as *soil*, its more acceptable name. In these investigations, students have opportunities to explore what goes together to make soil. To help them understand that soil is a mixture of earth materials, they first observe and manipulate the individual ingredients: clay, silt, sand, gravel, pebbles, and humus. Then they put the ingredients together themselves to create homemade soil.

Through systematic encounters with the basic macroscopic components of soil, students start to understand the importance of this valuable and limited natural resource. But be aware that full appreciation is some years off. It is reasonable to expect that, after a week or more of work with soil and its components, putting model soil together, and analyzing samples collected from home or field, when asked what is necessary to plant some seeds, most primary students will respond enthusiastically, "Dirt!" But don't be dismayed. The seeds of understanding have been planted, and students will be far better prepared to increase their understanding of soil the next time the subject comes around.

As with all science experiences for primary students, the learning is richer when a connection is made between the classroom and the real world. In the case of soil studies, that means going to see soil where it naturally is—outdoors. Once you have taken students out to extend their studies to the environment, the implicit message is that it is OK to inquire into and explore such subjects at any time and any place. Soil investigations can provide one avenue for exposing students to the wonderful complexity of the world and for introducing some of the universal constants. There truly is soil everywhere on Earth. It's not always the same, but it is there.

Of course, when you go outdoors with students, they will not regard soil in these lofty interconnected ways. They will want to dig holes, make mud balls, and write on the sidewalk with a dirt clod. Encourage their questions, promote their discoveries as the most important truths in the world, and tell yourself they are growing closer to the earth and wiser about their relationship to it.

During their analysis of local soil samples, students cover small samples with water and shake them vigorously in order to separate and identify the soil's component ingredients. This experience exposes students to the idea that moving water can disrupt the structure of soil, breaking it down, and efficiently carrying away the smaller particles in the soil. This represents the basis phenomena of weathering and erosion, two fundamental processes that alter and shape Earth's surface.

It is important for students to learn about the configurations and distribution (sources) of one particularly important earth material: water. It is important to know that most of Earth's water is in the ocean, and that only a small portion is fresh water distributed rather unevenly over Earth's surface as gas, liquid, and solid. Solid water in glaciers and ice caps is largely inaccessible to life, as is the water vapor suspended in the atmosphere. Accessible water for life, including humans, is found in lakes, ponds, rivers, and streams, and reasonably close below Earth's surface as groundwater delivered through springs and wells. As students continue through the grades, they will come to understand that one of the great scientific, social/political, and engineering challenges facing humans in their lifetime will be achieving universal access to high-quality fresh water for all people.

By working with a variety of earth materials and using those materials to study soil, students come to understand that different earth materials make different kinds of soil. They also work with engineering designs to prevent erosion and ways to represent land and water. The experiences students have in this investigation contribute to the disciplinary core ideas **ESS1.C, The history of planet Earth; ESS2.A, Earth materials and systems; ESS2.B, Plate tectonics and large-scale system interactions; ESS2.C, The roles of water in Earth's surface processes;** and the core ideas of **engineering design (ETS1.A, ETS1.B,** and **ETS1.C)**.

Engaging in Science and Engineering Practices (SEP)

In this investigation, students engage in these practices.

- **Asking questions** about how best to take soil apart to find out its component parts to compare one sample to another. **Defining a simple problem** about how to use tools to develop a process to study soil.
- **Developing and using models** to represent the shape and kinds of land and water; compare the models to identify common features and differences in photographs, drawings, and maps; and distinguish between the actual land and formations and the model. Use the models to identify the relationships between the land formations and water.

NGSS Foundation Box for SEP

- **Define a simple problem** that can be solved through the development of a new or improved object or tool.
- **Ask questions** based on observations to find more information about the natural world.
- **Distinguish between a model** and the actual object, process, and/or events the model represents.
- **Compare models** to identify common features and differences.
- **Develop and/or use a model** to represent amounts, relationships, relative scales, and/or patterns in the natural world.
- **Plan and conduct an investigation** collaboratively to produce data to serve as the basis for evidence to answer a question.
- **Make observations** (firsthand) to collect data that can be used to make comparisons.
- **Make predictions** based on prior experiences.
- **Record information** (observations, thoughts, and ideas).
- **Use and share pictures, drawings,** and/or writings of observations.
- **Use observations** (firsthand or from media) to describe patterns and/or relationships in the natural world in order to answer scientific questions.
- **Make observations** (firsthand or from media) to construct an evidence-based account for natural phenomena.
- **Compare multiple solutions** to a problem.
- **Construct an argument** with evidence to support a claim.

INVESTIGATION 4 — *Soil and Water*

NGSS Foundation Box for SEP

- **Obtain information** using various texts, text features, and other media that will be useful in answering a scientific question.
- **Communicate information** or design ideas and/or solutions with others in oral and/or written forms using models, drawings, writing, or numbers that provide detail about scientific ideas, practices, and/or design ideas.

NGSS Foundation Box for CC

- **Cause and effect:** Events have causes that generate observable patterns; simple tests can be designed to gather evidence to support or refute student ideas about causes.
- **Scale, proportion, and quantity:** Relative scales allow objects and events to be compared and described (bigger and smaller; hotter and colder; faster and slower).
- **Stability and change:** Some things stay the same while other things change. Things may change slowly or rapidly.

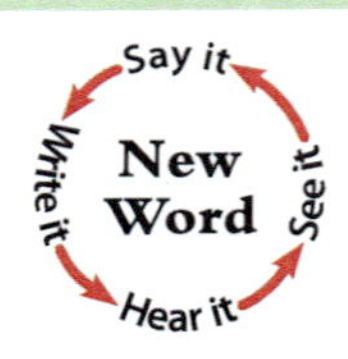

Decay
Fresh water
Gas
Humus
Lake
Liquid
Ocean
Pond
Retain
River
Salt water
Soil
Solid
Stream

- **Planning and carrying out investigations** with soil samples, both homemade and locally collected, in order to use tools to make comparisons.
- **Analyzing and interpreting data** dealing with different soil samples, and make predictions based on previous work with other earth materials. Record information in a notebook using drawings and words and share those observations with partners to answer questions.
- **Constructing explanations and designing solutions** about soil components through firsthand investigations, text, and video. Compare multiple solutions to preventing or slowing erosion.
- **Engaging in argument from evidence** to answer the question of what is soil and the claim that soil doesn't change.
- **Obtaining, evaluating, and communicating information** about soil, natural sources of water, soil erosion by wind and water, and how to represent land and water formations using media to gather information to answer questions and by communicating information orally and in written forms using models, drawings, words, and numbers.

Exposing Crosscutting Concepts (CC)

In this investigation, the focus is on these crosscutting concepts.

- **Cause and effect.** Wind and water can change the shape of land. People can do things to slow or prevent soil erosion by wind and water. Soil conditions impact things living in the soil, and things living in the soil impact the soil conditions.
- **Scale, proportion, and quantity.** Models of land and water show relative scales so that features can be described and compared.
- **Stability and change.** Erosion by wind and water can happen slowly or quickly depending on the circumstances.

Connections to the Nature of Science

- **Science addresses questions about the natural and material world.** Scientists study the natural and material world.

Connections to Science, Technology, Society, and the Environment

- **Influence of engineering, technology, and science on society and the natural world.** Every human-made product is designed by applying some knowledge of the natural world and is built by using natural materials.

Conceptual Flow

Soil and water as earth materials and natural resources are the phenomena of this investigation. The guiding question is how can we apply what we know about the ways land and water interact?

The **conceptual flow** for this investigation starts with **soil, an earth material**. Soil is the layer of **rock of different sizes** mixed with **humus, decomposing organic matter** that covers much of Earth's solid surface. Students make soil, using humus, sand, gravel, and pebbles. They use geologist tools—dry shaking, screening, and mixing with water—to separate its components

In Part 2, students investigate schoolyard soils from different locations to compare their properties (color, texture). Students plan which tools to use to study the soil.

Students have been using water in their study of solid earth materials and now turn their attention to **water as an earth materials and natural resource**. In Part 3, students learn about natural **sources of water**, both **fresh water** (rivers, streams, ponds, lakes) and **salt water** (oceans, bays, salt marshes).

In Part 4, students obtain information using text and photos to compare engineering designs to **prevent erosion** by wind and water. Students are introduced to **ways to represent the shape and kinds of land and water in an area** using photographs, drawings, and maps. Students use these models to identify relationships and common features and differences.

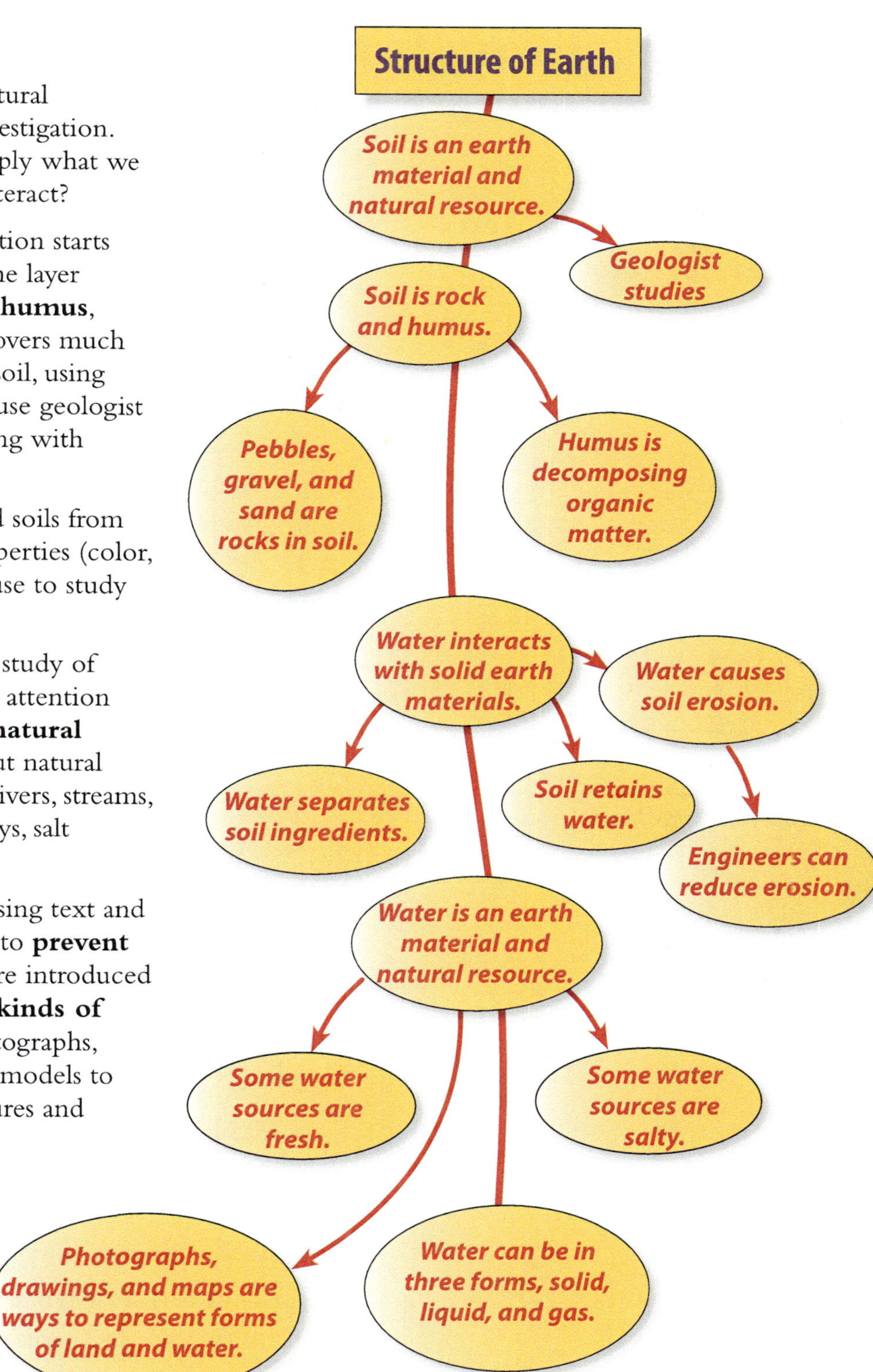

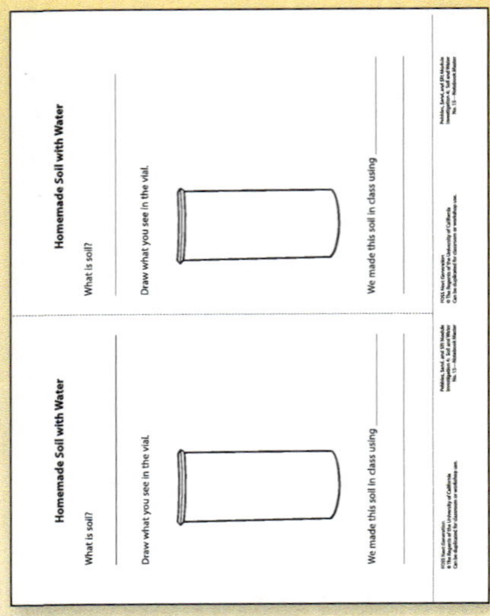

No. 15—Notebook Master

MATERIALS *for*

Part 1: *Homemade Soil*

For each student

- 1 Vial with cap
- 1 Paper plate
- 1 Self-stick note
- 1 Hand lens
- ❑ 1 Notebook sheet 15, *Homemade Soil with Water*

For each pair of students

- 1 Set of three screens
- 4 Containers, 1/4 L
- 1 Plastic cup
- 1 Zip bag, 1 L

For the class

- 1 Basin
- 1 Metal spoon
- • Potting soil
- 1 Vial
- 8 Vial holders
- 4 Plastic cups
- • Paper towels ★
- 1 Zip bag, 4 L
- • Transparent tape ★
- • Water ★
- • Sand, unwashed
- • Gravel
- • Small pebbles
- 1 Whisk broom and dustpan
- 1 Pitcher
- 1 Projection system (optional) ★

For embedded assessment

- ❑ • *Performance Assessment Checklist*

★ Supplied by the teacher.   ❑ Use the duplication master to make copies.

GETTING READY *for*

Part 1: *Homemade Soil*

1. Schedule the investigation

This part will take three active investigation sessions. Sessions 1 and 2 could be done together in one long period. Session 3 happens the day after students set up their soil tests.

NOTE

To prepare for this investigation, view the teacher preparation video on FOSSweb.

2. Preview Part 1

Students put together and take apart soils. They are introduced to humus, an important soil ingredient. They mix together homemade soil containing sand, gravel, pebbles, and humus. They shake some of the soil on a paper plate and observe what happens. They use screens to separate the homemade soil. They shake soil and water together in vials and draw what they observe. The focus question is **What is soil?**

3. Prepare potting soil

Potting soil is used as a source of humus (organic material) in this investigation. Scoop one vial full of potting soil into a zip bag for each pair of students. Save the zip bags at the end of this part.

4. Measure earth-material ingredients

In Step 4, you mix homemade soil as a demonstration. Measure out the ingredients in advance. Put the materials in four cups.

2 cups of sand, unwashed

1 cup of gravel

1 cup of small pebbles

NOTE

If you are not the first person to use the kit, check for a large zip bag labeled "Homemade Soil" from the previous use. Use that soil or make additional if needed.

5. Plan for vial labeling

Students will use the self-stick notes in the kit to label their vials. Have transparent tape at hand to secure labels to vials.

6. Locate a vial storage area

Each student will have one vial with the mixture of water and soil. You need to designate a location for storing the vials in holders, or have students keep them on their tables. These vials will be observed after 1 day and then kept intact for Part 2.

7. Plan for water

In session 2, fill a pitcher with water. Provide paper towels.

8. Plan assessment: performance assessment

Observe students as they separate the soil ingredients by shaking, screening, and setting up soil layering tests using water. Use these observations and 30-second interviews to assess students science and engineering practices. Carry the *Performance Assessment Checklist* with you to note progress.

FOCUS QUESTION

What is soil?

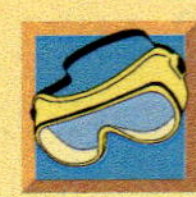

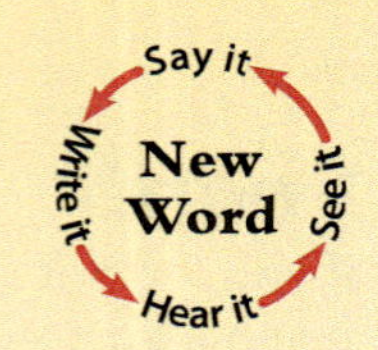

Materials for Steps 4–5

- *Basin*
- *Cups of gravel, small pebbles, and sand*
- *Metal spoons*
- *Student bags of humus*

NOTE

Reduce the quantity of ingredients if you are reusing homemade soil from a previous class use of the kit.

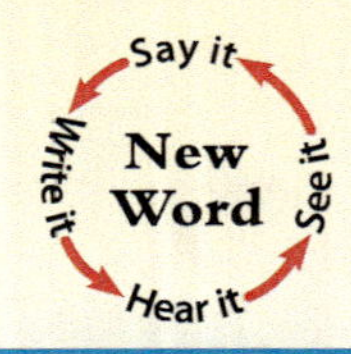

EL NOTE

Make a word web or diagram to show soil and its component parts—rock particles and humus.

GUIDING *the Investigation*

Part 1: *Homemade Soil*

1. Introduce a new material

Call students to the rug. Say,

We have been investigating rock and how it weathers into small pieces and is moved by water. But when we go into a natural area, we don't just see rock and water on the land. There is another important material on Earth's land surface for us to investigate.

Hold up a zip bag of potting soil. Explain that each pair will get a bag. They should first look at the material through the bag and then open the bag a bit to smell and touch the contents.

Demonstrate how to smell the soil by waving your hand over the material to bring the scent toward your nose. Caution students never to smell an unknown substance unless they are told that it is OK to do so by the teacher.

2. Distribute materials

While still at the rug, distribute a bag of humus to each pair and ask students to start examining the contents. Allow about 5 minutes for this observation.

3. Introduce *humus*

Ask students to describe the appearance, texture, and smell of the material. Tell them that the material in the bag is **humus** (HEW•mus). Humus is mostly plant and animal material that has **decayed** or rotted.

4. Describe making soil

Without using the word *soil*, tell students that you have a recipe for an earth material that includes humus. Hold up the four cups containing sand, gravel, and small pebbles one at a time, name the contents, and dump them into the basin. Explain that the last ingredient, the humus, will be added by them.

5. Add the humus

Have a few students at a time add their humus to the mixing basin. When all of the humus is added, stir the mixture. Save the zip bags for later soil investigations.

6. Introduce *soil*

Ask students to name the parts of the mixture. Then say,

The name for the mixture we just made is ***soil****. Soil is a combination of earth materials, like sand, gravel, and pebbles, and decaying plant material, like humus. Soil covers almost all of Earth's land surface.*

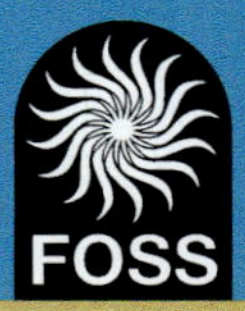

7. Introduce soil separating

Ask students to talk with a partner and share what they know about soil and what questions they have.

Tell students that their job now is to find out how to take soil apart. Suggest that they start by placing a sample of the soil on a paper plate. They should examine the soil, try to separate it by hand, and shake the plate back and forth as they did with the sand.

8. Distribute plates and soil samples

Have students move to their tables. Distribute two paper plates and a half cup of soil to each pair of students. Ask them to work together to divide the soil fairly between the two paper plates.

9. Assess progress: performance assessment

Have students begin examining their soil samples. When they are ready to shake the sample, encourage them to keep the plates on the table as they shake. As they do Steps 10–19, conduct 30-second interviews and assess students' performance.

What to Look For

- *Students ask questions based on observations of the soil mixture and are able to define the problem: the need to separate the soil into different sized particles. (Asking questions and defining problems; ETS1.A: Defining and delimiting engineering problems.)*
- *Students test the screens and other methods to find a solution for separating the soil. (Planning and carrying out investigations; ETS1.B: Developing possible solutions; cause and effect.)*
- *Students can analyze their soil separation results and describe how they would improve their methods if they had to solve the problem again. (Analyzing and interpreting data; ETS1.C: Optimizing the design solution.)*

10. Discuss observations

Discuss their success at using the shake method to separate the soil. Review what they put into the soil (humus, sand, gravel, small pebbles). Encourage students to separate these ingredients.

11. Use screens to separate the soil

Ask students to suggest other ways to separate the soil. Talk about screens and review how to use them.

a. *Put a paper plate under a container.*

b. *Put the large-mesh screen on top of the container.*

c. *Shake the soil on the screen and save what doesn't go through.*

d. *Continue the process with the other screens.*

EL NOTE

Model the procedure without actually using the soil.

Materials for Step 8

- *Paper plates*
- *Cups of soil*

SCIENCE AND ENGINEERING PRACTICES

Asking questions and defining problems

Planning and carrying out investigations

Analyzing and interpreting data

DISCIPLINARY CORE IDEAS

ETS1.A: Defining and delimiting engineering problems

ETS1.B: Developing possible solutions

ETS1.C: Optimizing the design solution

CROSSCUTTING CONCEPTS

Cause and effect

Materials for Step 12

- *Screens*
- *Containers*
- *Paper plates*
- *Soil samples*

12. Start screening

Explain that each pair of students should combine their soil samples and work together with a set of three screens and four containers. Distribute materials and begin screening. Students should notice that some, but not all, of the humus goes through all of the screens.

13. Discuss screening

Call students to the rug. Discuss what happened with the screening. Ask,

➤ *What problem were you trying to solve and what did you do to solve it?*

➤ *Were you able to separate the soil with the screens?*

➤ *What parts of the soil could you separate out with the screens?* [Possibly sand, gravel, and small pebbles.]

➤ *Where did the humus end up?* [In each container.] *Why?*

➤ *What do the large pieces of humus look like? The small pieces?*

TEACHING NOTE

The instructions can be intentionally vague so you can assess how well students plan and perform the separation.

SCIENCE AND ENGINEERING PRACTICES

Planning and carrying out investigations

14. Plan soil and water investigation

Ask students to brainstorm other ideas about how to separate the parts of the soil. When water is suggested, provide this minimal guidance.

a. *Mix your soil parts together again. Clean and stack the containers and screens and bring them to the materials station.*

b. *Vials with caps are available at the materials station.*

c. *Set up what you think will make a good investigation.*

d. *Raise your hand when you are ready for water.*

e. *See if you can separate the parts of the soil using water.*

Materials for Steps 15–17

- *Vials and caps*
- *Pitcher of water*
- *Self-stick notes*
- *Transparent tape*
- *Vial holders*

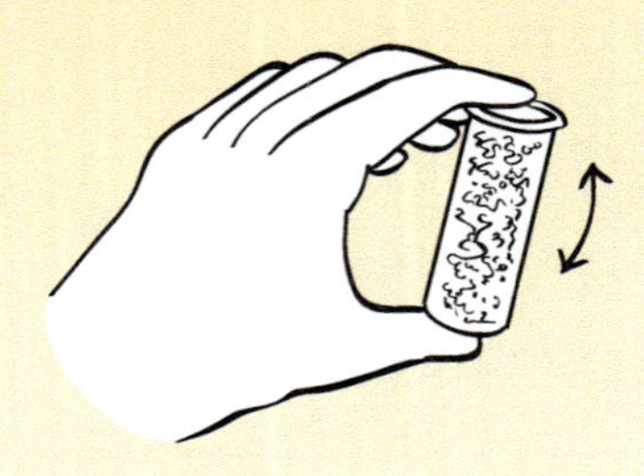

15. Start soil and water investigation

Let Getters get vials (without caps). There is a vial for each student. When they have soil samples in the vials, fill the vials with water.

16. Distribute caps and shake

Distribute the vial caps, so students can shake the vials. Ask students to describe what's happening to the contents as they shake the vials and when they let the vials rest. Remind students to hold the caps in place.

17. Let vials sit overnight

Tell students that they will let the vials sit overnight and observe them the next day. Ask them to predict what will happen. Have students label their vials with self-stick notes and place them in a vial holder. Have Getters bring the holders to the storage location.

18. Clean up

Have students recycle the leftover soil back into the basin. They should brush stray soil on their tables onto the paper plates and then into the basin. The broom and dustpan can be pressed into service to clean up larger spills. Have students bring their other materials to the materials station.

Save the remaining soil in the 4 L zip bag. Label it "homemade soil" for the next time the kit is used.

Materials for Step 18

- *Whisk broom and dustpan*
- *Zip bag, 4 L*
- *Paper towels*

BREAKPOINT

19. Review separation

At the rug, review the homemade soil and the methods students used to separate it. Remind students not to shake the vials.

20. Focus question: What is soil?

Ask the focus question and project or write it on the board.

➤ *What is soil?*

Explain that today they will observe what happened in their soil-and-water vials, and will draw a picture of what they see.

21. Observe and draw the vials

Have students move to their tables. Distribute the hand lenses and vials, reminding students not to shake the vials. Give each student notebook sheet 15, *Homemade Soil with Water*. Have them begin observing and drawing. Allow 10–15 minutes.

Materials for Step 21

- *Soil vials*
- *Hand lenses*
- ***Homemade Soil with Water*** *sheets*

22. Have a sense-making discussion

Ask them what they observed. As students describe a layer, write the word for that type of material on the board or word wall and reinforce the vocabulary. Projecting the notebook sheet can be useful during this discussion.

Students may describe their observations as humus (wood, bark, leaves), clay, silt, sand, gravel, and pebbles.

They may notice that some of the humus is floating on the top and some is mixed with the other materials.

Ask students to share how their thinking about soil has changed. If needed, provide prompts such as I used to think that soil _____ , but now I know that _____ .

SCIENCE AND ENGINEERING PRACTICES

Constructing explanations

23. Review vocabulary

Review key vocabulary added to the word wall in this part.

24. Answer the focus question

Have students glue the sheet into their notebooks. Have them answer the focus question on the top of the sheet or on the next blank page in their notebook.

25. Store the vials

Have students make sure their name labels are still on the vials. Store the vials in a secure location for Part 2, when students compare the homemade soil to soil collected outdoors.

WRAP-UP/WARM-UP

26. Share notebook entries

Conclude Part 1 or start Part 2 by having students share their notebook entries. Ask students to open their science notebooks to the most recent entry. Read the focus question together.

➤ *What is soil?*

Ask students to pair up to

- share their answers to the focus question;
- describe their labeled drawings.

To deepen student understanding, have them discuss the claim, "Soil does not change." Tell students that you heard someone offer this claim and write it on the board. Underneath make a T-Chart with "Evidence for" and "Evidence against." Tell students to talk with their group members to come up with ideas that might support this claim and ideas that refute this claim. [Students might provide evidence that soil changes because of the weathering of rocks, decaying of animals and plants, amount of moisture and air.]

Claim: Soil does not change.

Evidence for	Evidence against

SCIENCE AND ENGINEERING PRACTICES

Engaging in argument from evidence

CROSSCUTTING CONCEPTS

Stability and change

TEACHING NOTE

Students will revisit this claim and evidence at the end of the next part after they have more information.

ELA CONNECTION

This suggested strategy addresses the Common Core State Standards for ELA.

SL 1: Participate in collaborative conversations.

MATERIALS *for*

Part 2: *Local Soil*

For each pair of students

- 1 Plastic cup
- 1 Metal spoon
- 1 Small piece of scratch paper ★
- 2 Hand lenses
- 1 Self-stick note
- 1 Paper plate
- 1 Vial with cap
- 1 Set of screens
- 2 Vials of homemade soil (from Part 1)
- 4 Containers, 1/4 L
- ❑ 2 Notebook sheet 16, *Local Soil with Water*
- 2 *FOSS Science Resources: Pebbles, Sand, and Silt*
  - "What Is in Soil?"
  - "Testing Soil"

For the class

- 1 Whisk broom and dustpan
- 1 Bottle brush
- 1 Basin
- 8 Vial holders
- 1 Pitcher
- 1 Carrying bag ★
- • Water ★
- • Paper towels ★
- • Transparent tape ★
- 1 Big book, *FOSS Science Resources: Pebbles, Sand, and Silt*
- 1 Computer with Internet access ★

For embedded assessment

- ❑ • *Embedded Assessment Notes*

★ Supplied by the teacher. ❑ Use the duplication master to make copies.

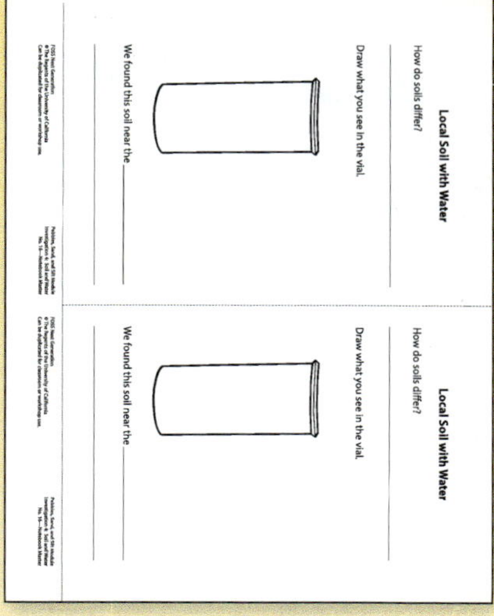
Local Soil with Water
How do soils differ?
Draw what you see in the vial.
We found this soil near the ______

Local Soil with Water
How do soils differ?
Draw what you see in the vial.
We found this soil near the ______

No. 16—Notebook Master

GETTING READY *for*

Part 2: *Local Soil*

> **TEACHING NOTE**
>
> *One option is to provide each group with a card describing a specific site for getting a soil sample. The cards then serve as labels for the samples.*

1. Schedule the investigation

This part will take five sessions, three for active investigations and two for readings. The first session is an outdoor search for local soil followed by a reading session. In session 3, students study the local soil, and the next day (session 4) they observe the layers of soil in water. The final session is for reading.

2. Preview Part 2

Students go on a schoolyard field trip to collect soil samples. They try to find soil in as many places as possible: next to sidewalks, near trees, and in landscaped areas. Students study their schoolyard soil samples. They shake vials with the soil and water, then draw the results. They compare the vials and drawings of their schoolyard samples with the vials and drawings of the homemade soil. The focus question is **How do soils differ?**

3. Plan for additional help

Schedule session 1 of this part for a day when you have volunteers (older students or adults) or aides to help you with the logistics.

4. Select the outdoor site

Take a walk around your schoolyard to look for possible soil-collecting sites: next to sidewalks, in sidewalk cracks, near planted areas and trees, and in grassy areas such as sports fields. Each pair of students will collect about one-half cup of soil.

5. Get permission to dig in the schoolyard

Tell the principal and other personnel who should know that you will be taking groups of students out to collect soil samples in the schoolyard. Check to see if there are any places students should *not* dig.

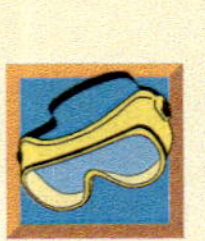

6. Check the site

Tour the outdoor site on the morning of an outdoor activity. Check for unsightly and distracting items where the students will be digging.

7. Designate a storage location

Each pair of students will produce one soil-and-water vial. Plan a location for storing the vials and the cups with the remaining soil samples.

8. Plan for vial labeling

Students will use self-stick notes to label their vials. They may need a piece of transparent tape to secure the label to the vial.

9. Plan for cleanup

Don't let the soil sit in the vials too long after this part. The organic material in the soil can begin to smell. You can reuse the soil by placing it in a basin to use later for planting activities. (See the Interdisciplinary Extensions.)

10. Preview the video

Preview the video, *All about Soil* (duration 21 minutes). The video provides a good review of concepts students have learned in previous investigations dealing with the breaking of large rocks to form smaller pieces of rock through the process of weathering, and the movement of those pieces to new locations through the process of erosion. New information includes a description of soil layers (top soil, subsoil, bottom, and bedrock), the concept of parent rock—the original rock that breaks into pieces—types of soil (sandy, clay, loam) and what makes for good topsoil. Soil as a limited natural resource that takes a long time to form and ways to conserve soil are discussed. These topics are included as chapters.

- Introduction
- Investigation: How does soil form?
- Layers of soil
- Properties of soil
- Compost
- Climate and soil
- Soil conservation

The link to this video for teachers is in the Resources by Investigation on FOSSweb.

11. Plan to read *Science Resources*: "What Is in Soil?" and "Testing Soil"

Plan to read "What Is in Soil?" during a reading period after the outdoor soil search. Plan to read "Testing Soil" after students have completed their study of the local soil samples.

12. Plan assessment: notebook entry

In Step 18 of Guiding the Investigation, students draw and label what they observe in the settled vial of soil and water. Look for students' ability to record observations of soil layers in water and to compare two soil samples.

FOCUS QUESTION

How do soils differ?

TEACHING NOTE

Each soil sample should be from one place only—no combined samples. Encourage students to see if anything is growing in the soil they are collecting.

Materials for Steps 4–5

- *Cups*
- *Metal spoons*
- *Scratch paper*
- *Tape*
- *Carrying bag*

SCIENCE AND ENGINEERING PRACTICES

Planning and carrying out investigations

TEACHING NOTE

Have students store the cups until the next day, when they will be studied and compared again. Any animals that students find in the soil should be returned to the outdoors.

GUIDING *the Investigation*

Part 2: *Local Soil*

1. Introduce the soil search

Call students to the rug. Review with them the ingredients they put together to make soil. Explain that soil can be found in many places outdoors. Pause for students to think about where they would find soil. Call on a few volunteers to share. Offer that one clue for finding soil is to look for places where plants are growing.

2. Focus question: How do soils differ?

Ask the focus question and project or write it on the board. Brainstorm ideas and write them on the board.

➤ *How do soils differ?*

3. Explain the ground rules

Ask students what they could do to find out. When a student mentions collecting soil, suggest they make a plan to investigate soil from different locations around the school. Each pair of students will get one metal spoon and a plastic cup. Each pair is to use the spoon to dig up a sample of soil from only one location. They need to fill the cup no more than halfway with their samples. The pairs should stay together and decide where to dig up soil. Students will use the soil samples to answer the focus question.

4. Go outdoors

Take students and the materials outdoors. Distribute the cups and spoons when students are ready to start digging.

5. Label the soil samples

When each pair has dug up a sample, help students label their cups with a small piece of paper, including their names and where they found the soil. The label can be put in the cup on top of the soil sample unless the soil is wet, in which case the label should be taped to the cup. Do this outdoors so that students don't forget where they found their samples.

6. Return to class

Back in the classroom, have students put the cups of soil near each other on a table. Have students share observations by asking,

➤ *Are all the soils the same texture?*

➤ *Are all the soils the same color? Let's line them up from the darkest soil to the lightest soil by color.*

➤ *Which two soil samples look very different? Where were they collected?*

➤ *In which soils could you see things growing?*

READING *in Science Resources*

7. Read "What Is in Soil?"

Remind students that in the last part they were exploring the question, "What is soil?" Point out the title and explain that this article will review what they have been learning and provide more information about soil. Give students time to preview text by looking at and discussing the photographs with a partner. Tell them to review the words in bold to make sure they know their meaning. Make sure students understand that nutrients help plants grow but are not the "food" for plants.

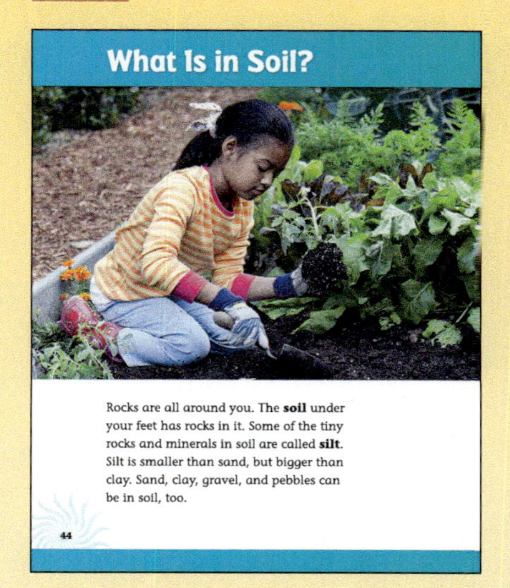
What Is in Soil?

Rocks are all around you. The **soil** under your feet has rocks in it. Some of the tiny rocks and minerals in soil are called **silt**. Silt is smaller than sand, but bigger than clay. Sand, clay, gravel, and pebbles can be in soil, too.

44

Read the story aloud or have students read independently. After each page, tell students to pause and take turns summarizing the main idea.

8. Have a sense-making discussion

Discuss the differences students see between the soils on the last page of the article. Use the table below as a guide.

Soil on Left	Soil on Right
Layer of humus on top	Layer of humus on top
Rich soil supporting plant life—roots big and deep	Small roots
Small particles of rock—clay, sand, silt	Many different sizes of rocks—cobbles, pebbles, sand
Different color layers	Unclear layers with same color

Then ask,

- ➤ *How do these soils compare with the soils we studied in class?*
- ➤ *What did you already know about soil?*
- ➤ *What was new information for you?*

Read the text aloud modeling strategies that support reading comprehension. Pause at the word *retain* and model how to determine the meaning. [The text doesn't provide enough context; it could mean to hold or to drain water. Best to look it up in the glossary.]

ELA CONNECTION

These suggested strategies address the Common Core State Standards for ELA.

RI 1: Ask and answer questions to demonstrate understanding.

RI 2: Identify the main topic of the text.

L 4: Determine or clarify the meaning of unknown or multiple-meaning words and phrases.

Pose a few of the following questions at a time for students to discuss in their groups and then share with the class. Encourage students to support their responses with evidence from the text.

- *How are worms good for the soil?* [They mix and break up soil.]
- *What animals besides worms live in the soil?* [Ants, moles, gophers, etc.]
- *Why do these animals live in the soil?* [Protection, access to food (roots of plants), etc.]

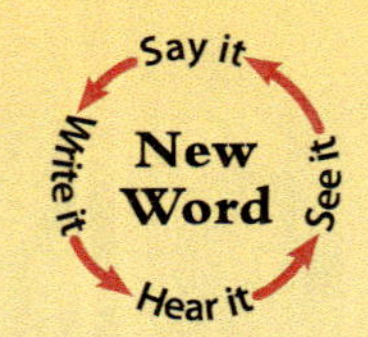

Ask students to think about and discuss these questions about rain water and soil.

- *What happens to the soil when it rains? Where does the water go?* [Some of the water soaks into the soil.]
- *Will different kinds of soils **retain** (or hold) water in the same way?*
- *How might sandy soil retain water?*
- *How might soil with lots of humus retain water?*

CROSSCUTTING CONCEPTS

Cause and effect

Next, ask students to think about and discuss the following questions about human use of soil. Write the questions on the board or chart paper and give students time to mull them over. Pass out self-stick notes for them to write their ideas and then post them on the chart paper. Group similar ideas.

- *Why is soil important?* [Plants grow in soil.]
- *Plants are resources people use for many things. What do people use plants for?* [Food, shade, lumber, paper, fires (fuel).]

SCIENCE AND ENGINEERING PRACTICES

Obtaining, evaluating, and communicating information

9. Summarize the reading

One option is to have students work together in small groups to create a poster to communicate this summary information. Write the questions on the board and have students come up with visuals and labels to show different types of soil and the role of humus and worms.

- *What is humus?* [Decayed dead plants and animals.]
- *What does humus do for soil?* [Provides nutrients and helps soil retain water.]
- *How are worms good for soil?* [Worms mix and turn soil, and they break it apart.]
- *Describe different kinds of soil.* [Some soil has more clay, some more sand, some more pebbles and gravel, some more humus.]

ELA CONNECTION

This suggested strategy addresses the Common Core State Standards for ELA.

W 7: Participate in shared research and writing projects.

BREAKPOINT

10. Discuss the procedure

Call students to the rug and review the field trip to collect soil samples. Ask,

➤ *What can we do to observe the soil to answer the focus question, how do soils differ?*

Listen to and discuss students' ideas, providing an opportunity for students to ask questions, make predictions, and to define a simple problem about how to use tools to develop a process to study soil.

Tell them that you will distribute the soil samples to each pair. Each pair will get a paper plate on which to pour the sample for observing. After this first observation students will report their discoveries. Encourage them to feel the texture of the soil.

SCIENCE AND ENGINEERING PRACTICES

Asking questions and defining problems

11. Distribute materials and make soil observations

Have students move to their tables. Distribute paper plates, hand lenses, and the soil samples from the field trip. Encourage students to work over their paper plates to help contain spills. Allow 5–10 minutes for observing.

Materials for Step 11

- *Cups of soil samples*
- *Paper plates*
- *Hand lenses*

12. Review observations

Call for attention and have students report their discoveries. Ask them to tell where they found their soil as well as what they found in the sample.

SCIENCE AND ENGINEERING PRACTICES

Planning and carrying out investigations

Analyzing and interpreting data

13. Review separating soil

Ask students to predict what they would find if they separated their soil samples. Review the several techniques used to separate soil, and ask which technique did the best job of separating the soil. [Screening, shaking the soil with water and letting it settle.]

14. Describe available tools

Tell students that they can use the screens and containers to separate the parts of the soil and they can use a vial with water. Tell them that each pair of students can get a set of three screens and a vial with a cap. Let the pairs plan and conduct their own soil-separation investigations. Make the materials available at the materials station.

Materials for Steps 14–16

- *Screens*
- *Containers*
- *Vials and caps*
- *Pitcher of water*
- *Self-stick notes*
- *Vial holders*

15. Start vial procedure

Have students raise their hands when they are ready for water in their vials. As they observe the water and soil in the vials and then shake them, ask them to describe what they observe and compare it to what they remember happening with homemade soil in Part 1.

TEACHING NOTE

Remind students to hold the caps in place with their thumbs.

Distribute a self-stick note to each pair; have them put both students' names on the label. Allow about 10 minutes for filling, shaking, and observing the vials.

16. Let vials sit overnight

Have students place the vials and the plastic cups with dry soil next to each other in the storage area. Have the Getters return the paper plates, screens, containers, and hand lenses. Ask students to make predictions about what they will see the next day.

BREAKPOINT

17. Describe notebook sheet

Call students to the rug and review. Ask them what question they were trying to answer and what they did to answer that question. Tell them that they should observe what happened to the soil and water overnight, and draw a picture of the vials to compare to other soils. They should answer the focus question.

Hold up a copy notebook sheet 16, *Local Soil with Water.* Point out the line under the vial where students should write where they found their soil samples. They should find this information on the label in their sample cups.

Materials for Steps 18–19

- *Vials of soil samples*
- *Cups of soil samples*
- ***Local Soil with Water*** *sheets*
- *Vials of homemade soil*

18. Observe and draw the soil samples

Distribute the vials and plastic cups to pairs of students. Ask them to observe the vials and draw what they see. They should compare their samples and drawings with another pair of students.

19. Answer the focus questions

Distribute the vials with the homemade soil and water from Part 1. Ask students to compare the two soil vials and the two drawings. Ask them to think of one property that is the same in the two soils and one property that is different, and answer the focus question.

20. Review vocabulary

Ask students to name the earth materials they found in their homemade soil. Record them on the board. Ask students for the earth materials they found in their schoolyard soil samples. Record them on the board. Remind students that soil is a combination of materials such as those on the board.

HOMEMADE SOIL
humus
sand
gravel
pebbles
SOIL SAMPLES
humus
silt
sand
gravel
pebbles

Ask,

➤ *How is our homemade soil like the soil samples we collected?* [They have some of the same materials.]

➤ *How is it different?* [Different soils can have different amounts and combinations of materials.]

TEACHING NOTE

Have students review the sand and clay drawings from Investigation 2 and compare them to the soil drawings.

21. Assess progress: notebook entry

Use notebook sheet 16, *Local Soil with Water*, to assess students' abilities to make and record observations.

What to Look For

- *Students record data accurately, including labels on their drawings.*
- *Students record observations that soils have different properties (color, texture, or different amounts of sand, gravel, pebbles, and humus).*

SCIENCE AND ENGINEERING PRACTICES

Constructing explanations

22. Clean up

Have students bring the cups with soil samples and the vials to the materials station. Remind students not to dump the soil into the sink drain. Shake up the vials briskly before dumping the contents into a basin. Dump soil remaining in the cups into the basin as well. Allow the soil to dry for later use, or put it back outdoors.

Have several students help rinse out the vials, caps, and cups in a basin of clean water. Set the vials, caps, and cups out to dry. The spoons used to dig the soil should be rinsed and dried too. The rinse water should be recycled outdoors.

Materials for Step 22

- *Basins*
- *Bottle brush*
- *Paper towels*

23. View the video

Refer students to the pictures in the article "What Is Soil?" and ask them if they think that soil forms slowly or quickly.

Tell students the video will tell more about how soil forms and different kinds of soil. Show the 21-minute video, *All about Soil*. List new words on the word wall as they are discussed in the video.

There are two demonstrations in the model that the students can relate to based on their previous experience. In one demonstration, rocks are placed in a jar with water and shaken for 6 minutes. The result is similar to what students saw when they washed their river rocks after rubbing them together.

The second demonstration answers the question, why are the larger pieces of rock found in the bottom layers of soil? This demonstration is similar to the settling out of rock particles in water with the larger particles on the bottom. See if students can make these connections between demonstrations in the video and experiences they have had in the investigations.

Testing Soil

Do plants grow better in soil or sand? Here's what you can do to find out.

1. Get four cups that are all the same size.
2. Fill two cups with potting soil that has lots of humus. Fill the other two cups with sand.
3. Plant three sunflower seeds in each cup.
4. Put the same amount of water in each cup.
5. Keep the cups in a sunny window, and record what happens.

48

TEACHING NOTE

See the Science Extensions in the Interdisciplinary Extensions at the end of this investigation for related student projects.

ELA CONNECTION

This suggested strategy addresses the Common Core State Standards for ELA.

RI 6: Identify the main purpose of the text.

SCIENCE AND ENGINEERING PRACTICES

Obtaining, evaluating, and communicating information

CROSSCUTTING CONCEPTS

Cause and effect

READING *in Science Resources*

24. Read "Testing Soil"

"Testing Soil" proposes an experiment that tests the importance of humus in soil and introduces students to the elements of experimental design.

Ask,

➤ *Do plants grow better in soil or sand?*

Briefly brainstorm and discuss answers. Ask students how they would investigate that question. Brainstorm and discuss students' ideas. Introduce the article as an example of how students could investigate the question.

Read aloud or have students read independently. In their notebooks have students draw a diagram for each numbered step in the procedure.

25. Discuss the reading

Before discussing the questions on page 49, be sure students have a clear understanding of the experiment. Ask,

➤ *What conditions are the same for both plants?* [Cup, seeds, water, location, and Sun exposure.]

➤ *What is different?* [What the plants are growing in, soil or sand.]

Then discuss the questions on page 49 of the reading.

➤ *Is this a good way to test the question?* [Yes, it is a fair test to see if soil or sand is better for plant growth. Everything is the same except what the plant is growing in.]

➤ *Two students planted seeds in soil and sand. Look at the plants above. Which seeds grew better? Why do you think that happened?* [The plants in the soil seemed to be bigger and greener. The soil had more nutrients. The soil might have retained more water and kept the plants moist. Maybe the plants in the sand dried out.]

You can restate the question "Which seeds grew better" by asking

➤ *What was the effect of using soil to grow plants?*

➤ *What was the effect of using sand to grow plants?*

And then the final question,

➤ *Do the test yourself. Draw or write about your results.*

WRAP-UP/WARM-UP

26. Share notebook entries

Conclude Part 2 or start Part 3 by sharing notebook entries. Ask students to open their science notebooks to the most recent entry. Read the focus question together.

➤ *How do soils differ?*

Ask students to pair up to

- share their answers to the focus question;
- describe their labeled drawings.

Refer students to the chart of evidence for and against the claim that soil doesn't change they made during the last Wrap-Up/ Warm-Up session. Ask students if they have changed their thinking and have more evidence to counter or support this claim. [Students should offer their observations of different types of soil found in the schoolyard and new information about how worms change the soil, in addition to weathering and decaying animals and plants.]

SCIENCE AND ENGINEERING PRACTICES

Engaging in argument from evidence

ELA CONNECTION

This suggested strategy addresses the Common Core State Standards for ELA.

SL 1: Participate in collaborative conversations.

TEACHING NOTE

*See the **Home/School Connection** for Investigation 4 at the end of the Interdisciplinary Extensions section. This is a good time to send it home with students.*

No. 17—Notebook Master

MATERIALS *for*

Part 3: *Natural Sources of Water*

For each student

- ❑ 1 Notebook sheet 17, *Water in Our Community*
- ❑ 1 Notebook sheet 18, *States of Water*
- 1 *FOSS Science Resources: Pebbles, Sand, and Silt*
  - "Where Is Water Found?"
  - "States of Water"

For each group of students

- 1 Set of Sources of Water cards

For the class

- 1 *Natural Sources of Water* poster
- 1 Big book, *FOSS Science Resources: Pebbles, Sand, and Silt*

For benchmark assessment

- ❑ • *Embedded Assessment Notes*

❑ Use the duplication master to make copies.

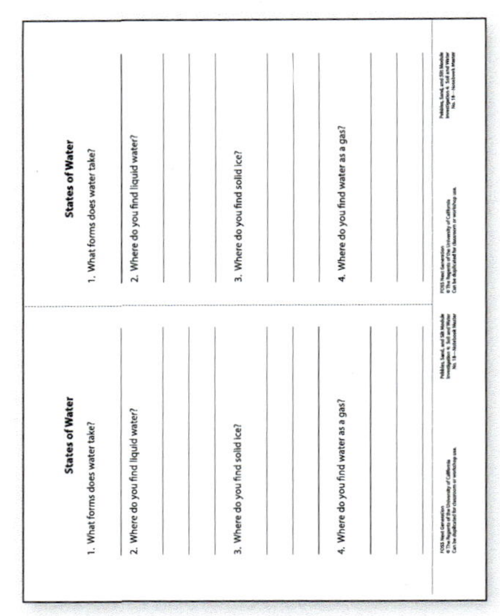
States of Water

1. What forms does water take?
2. Where do you find liquid water?
3. Where do you find solid ice?
4. Where do you find water as a gas?

States of Water

1. What forms does water take?
2. Where do you find liquid water?
3. Where do you find solid ice?
4. Where do you find water as a gas?

No. 18—Notebook Master

GETTING READY *for*
Part 3: *Natural Sources of Water*

1. Schedule the investigation
Part 3 will take two reading sessions.

2. Preview Part 3
Students read about sources of natural water, sort images of water sources, both fresh and salt, and discuss where water is found in their community. The focus question is **Where is water found in our community?**

3. Review the cards and poster
Inventory the Sources of Water card sets to make sure that all 18 cards are in each set. There are eight sets of cards in the kit. One set of Sources of Water cards contains two cards each of these nine sources: coral reefs, glaciers and ice, mangrove forests, ocean and seas, ponds and lakes, rocky coasts, salt marshes, sandy beaches, and streams and rivers.

Locate the *Natural Sources of Water* poster (shown in the sidebar) and plan where to display it in the classroom.

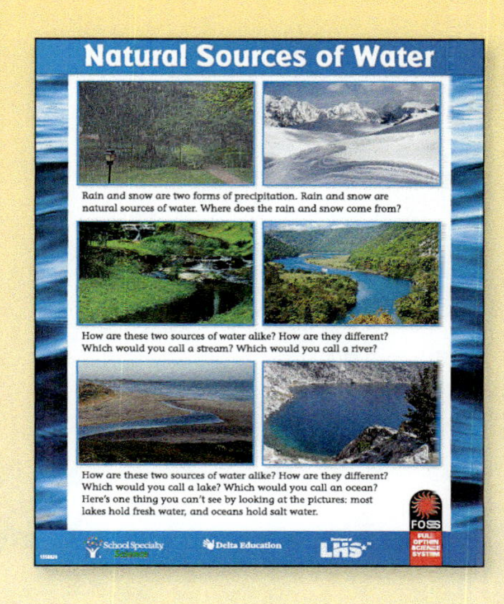

4. Plan to read *Science Resources*: "Where Is Water Found?" and "States of Water"
Plan to read "Where Is Water Found?" and "States of Water" during this part.

5. Plan assessment: notebook entries
In Step 9, students complete notebook sheet 18, *States of Water*, and in Step 11, students complete notebook sheet 17, *Water in Our Community*. After class, check students notebooks for evidence that students can identify the different states of water and can identify water sources in the local community.

FOCUS QUESTION

Where is water found in our community?

GUIDING *the Investigation*

Part 3: *Natural Sources of Water*

1. Focus question: Where is water found in our community?

Call students to the rug. Introduce the focus question by saying,

We have been using water to separate earth materials and learning about how water can move rocks and soil. I'm wondering how water affects landforms in our community. Let's start with where we find water.

Ask the focus question and project or write it on the board.

➤ *Where is water found in our community?*

Listen to and discuss students' ideas. Keep a list of the places on the board. Tell them that you will read to them so they can get more information about where water is found.

READING *in Science Resources*

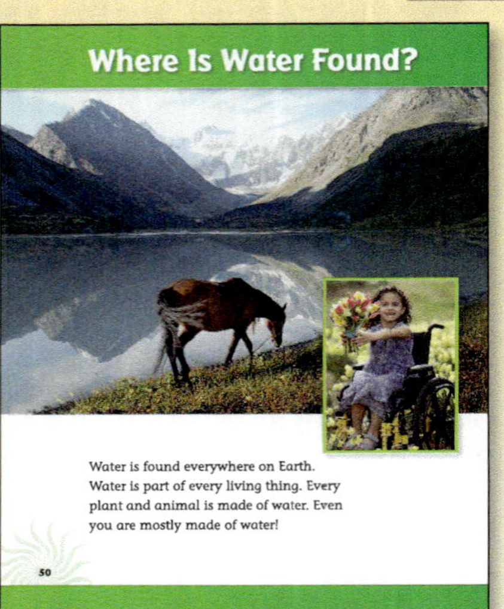
Where Is Water Found?

Water is found everywhere on Earth. Water is part of every living thing. Every plant and animal is made of water. Even you are mostly made of water!

50

2. Read "Where Is Water Found?"

Ask students to listen for ideas to add to the brainstorming list. Read the article aloud. Pause to discuss key points.

page 51	*Where are there creeks, **streams**, or **rivers** near us?*
page 52	*Where are there **ponds** or **lakes** near us?*
page 54	*How else do we use water?*
page 55	*Is there a saltwater **ocean** or sea near us?*
page 56	*Is there a salt marsh near us?*

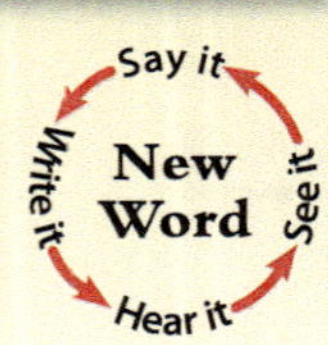

3. Discuss the reading

After the reading, ask,

➤ *Why is **fresh water** so important?* [People, animals, and plants must have fresh water to live.]

➤ *Most of the water on Earth is **salt water**. Is salt water important to us? Why do you think so?* [Many plants and animals live in the ocean. People use resources from the ocean as food.]

➤ *Where is fresh water found in our community?*

➤ *Where is salt water found in our community?*

Materials for Step 4
- *Sources of Water cards*

4. **Distribute the Sources of Water cards**
Shuffle the 36 cards in two sets of Sources of Water cards and distribute one card to each student. Students should study their own cards and then compare them with the card of the person sitting next to them. Pairs of students should describe how the water sources are the same and how they are different. Ask students to describe the shapes and kinds of land and water shown on the card.

5. **Read the article again**
Tell students you are going to reread the article. After you read about a water source, you will ask students to hold up their card if it describes that source.

Stop reading at the end of the first paragraph on page 51. Ask students who have cards showing streams or rivers to hold their cards up for all to see. Continue, sharing cards after reading about each source of water.

At the end of the article, the students holding the Glaciers and Ice cards will not have had a chance to stand up. Ask them to stand up now and describe the image on their cards. Ask students where glaciers and ice might be found on Earth. Depending on students' experience, they may report the top of high mountains. Most of the glaciers on Earth are in Antarctica and Greenland but they exist on nearly every continent, and there are many in the United States, most of them located in Alaska.

SCIENCE AND ENGINEERING PRACTICES

Obtaining, evaluating, and communicating information

6. **Sort the Sources of Water cards**
Give each group of students a set of the cards and let them sort them based on one property (fresh water or salt water; still water or moving water; rocks or no rocks).

READING *in Science Resources*

7. Read "States of Water"

The reading discusses forms of water (solid, liquid, and gas). Use it to review states of water.

Tell students that this article is about forms of water. Tell them to close their eyes and visualize water as a liquid. When they have their image in mind, have them turn to a neighbor and describe the liquid water. Next have students visualize water as a solid (ice or snow). Have them describe how it might feel and where they would find it. Ask students if they know another form of water [gas]. Tell students that they can't visual water as a gas because it's invisible.

Read aloud or have students read independently. After each section, pause and have students match the photographs that illustrate each of the ideas in the text. Have student discuss how the photographs help them understand the text.

States of Water

Liquid water is one state of water. We can pour it into a glass to drink. We spray it from a hose to water plants. Liquid water can drip from a fountain.

61

ELA CONNECTION

These suggested strategies address the Common Core State Standards for ELA.

RI 7: Explain how images contribute to and clarify text.

RF 4: Read with accuracy and fluency to support comprehension.

8. Discuss the reading

Review any questions students have about the reading and then refer them to the last page. Ask students to identify the three states of water. Reinforce the idea that water as a gas is water vapor in the air and that it is invisible. Clouds are made up of droplets of liquid water.

Have students share responses to these questions:

- *Where could we find water as a **liquid**?*
- *Where could we find water as a **solid**?*
- *Where could we find water as a **gas**?*
- *Do plants and animals need water to be in different forms—solid, liquid, gas?*

SCIENCE AND ENGINEERING PRACTICES

Obtaining, evaluating, and communicating information

9. Make a notebook entry

Hold up a copy of notebook sheet 18, *States of Water*. Read it to students and ask them to write or draw about water in each of its states or forms. Distribute a copy of the notebook sheet to each student.

Materials for Step 10
- *Natural Sources of Water* poster

fresh water
gas
lake
liquid
ocean
pond
river
salt water
solid
stream

10. Review vocabulary

Use the poster, *Natural Sources of Water,* to review what the students have read about and discussed. Have them look at the first two pictures and identify them as ways water falls to Earth from the sky: rain and snow.

Have students look at the rest of the pictures on the poster, then identify and describe the places that show where water gathers or is stored. Have them describe natural sources of water, including rain, snow, stream, rivers, lakes, and the ocean they have seen. Reinforce that water is a natural resource.

11. Answer the focus question

Hold up a copy of notebook sheet 17, *Water in our Community.* Read it to students and ask them to write and draw about a local source of water in their community. Ask them to show the shapes of the land and water in that area.

12. Assess progress: notebook entry

Collect students' notebooks after class and assess progress.

What to Look For

- *Students have obtained and can communicate information to identify sources of local water on the Water in Your Community notebook sheet.*
- *Students are able to identify the three states of water as shown on the States of Water notebook sheet.*

MATERIALS *for*

Part 4: *Land and Water*

For each student

1 *FOSS Science Resources: Pebbles, Sand, and Silt*

- "Erosion"
- "Ways to Represent Land and Water"

For the class

1 Big book, *FOSS Science Resources: Pebbles, Sand, and Silt*

For embedded assessment

❏ • *Embedded Assessment Notes*

For benchmark assessment

❏ • *Investigation 4 I-Check*

• *Assessment Record*

❏ Use the duplication master to make copies.

GETTING READY *for*

Part 4: *Land and Water*

1. Schedule the investigation

Part 4 will take two reading sessions and one session for the assessment.

2. Preview Part 4

Students compare a variety of solutions to slow down the effects of wind and water erosion on land. They go out on the schoolyard to look for erosion. They end the module by studying a variety of images representing different landforms and bodies of water and identify common features and differences. The focus question is **How can soil erosion be reduced?**

3. Plan to read *Science Resources*: "Erosion" and "Ways to Represent Land and Water"

Plan to read "Erosion" and "Ways to Represent Land and Water" during this part.

4. Plan assessment: notebook entries

In Step 4 of Guiding the Investigation, students answer the focus question. Check students' understanding of how engineers prevent erosion.

In Step 6, students draw a map of the schoolyard or park, including water sources and landforms. Check students' understanding about maps and representing land and water features.

5. Plan assessment: I-Check

Plan to give *Investigation 4 I-Check* at the end of the investigation. Read the items aloud to the whole class, and have students answer independently. Review students' responses using the What to Look For. Use assessment master 5, *Assessment Record,* to record students' responses.

GUIDING *the Investigation*

Part 4: *Land and Water*

FOCUS QUESTION

How can soil erosion be reduced?

Erosion

What happened to this road? People once drove on this road. During a big storm, waves crashed against the shore. They washed away the soil under the road. Parts of the road were destroyed.

68

READING *in Science Resources*

1. Read "Erosion"

Review the class chart made after reading the "Rocks Move" article and what students learned from the *All about Land Formations* video. Tell students that when water and wind move rocks it can sometimes cause problems. Have them think about and then share with a partner what they think those problems might be.

Review that erosion refers to the movement of rocks by water or wind. Tell students that this article shows examples of how erosion can cause problems and the different ways people go about solving these problems.

Tell students to look at the photograph on the first page and discuss what they observe. Ask,

➤ *What evidence is there of erosion?*

Next, have students look at the photograph on the fourth page and discuss what they think engineers are doing here to solve the problem of erosion on the riverbank. Point out the word "barrier" and ask students what they think it means.

Read aloud or have students read the first four pages independently. Review any confusing words or phrases or questions students have about the text. If students read independently, reread these pages aloud and then ask students to take turns telling a partner how engineers are solving the erosion problem on the riverbank.

Point out the photographs on the next two pages and have students discuss what they observe. Read the pages aloud and then have students reread the passages and match the photographs to the solutions described in the text.

Tell students that the next four pages show erosion by wind and ways to solve problems caused by wind erosion. Read aloud or have students read the rest of the article independently.

SCIENCE AND ENGINEERING PRACTICES

Constructing explanations and designing solutions

Obtaining, evaluating, and communicating information

ELA CONNECTION

These suggested strategies address the Common Core State Standards for ELA.

RI 3: Describe the connection between scientific ideas or concepts.

RI 5: Know and use text features.

SL 4: Recount an experience.

2. **Focus question: How can soil erosion be reduced?**
Ask the focus question and project or write it on the board.

➤ *How can soil erosion be reduced?*

3. **Have a sense-making discussion**
Go to the last page of the reading. Ask students to work in groups to compare the design solutions.

➤ *What kinds of materials are used in the solutions shown in the photographs?*

➤ *Why do you think those materials are being used?*

➤ *What features are the same?*

➤ *What features are different?*

➤ *What would a good design solution do?*

➤ *Which design do they think would work the best and why?*

4. **Answer the focus question**
Show students the focus question on the board and read it to them.

➤ *How can soil erosion be reduced?*

Ask students to turn to the next blank page in their notebooks and write the focus question. Ask them to use the information from the reading and from the class discussion to describe one way engineers reduce soil erosion. Encourage them to use photographs and words in their description.

POSSIBLE BREAKPOINT

CROSSCUTTING CONCEPTS

Cause and effect

NOTE

Go to FOSSweb for *Teacher Resources* and look for the Crosscutting Concepts—Grade 2 chapter for details on how to engage young students with this concept.

CROSSCUTTING CONCEPTS

Stability and change

READING *in Science Resources*

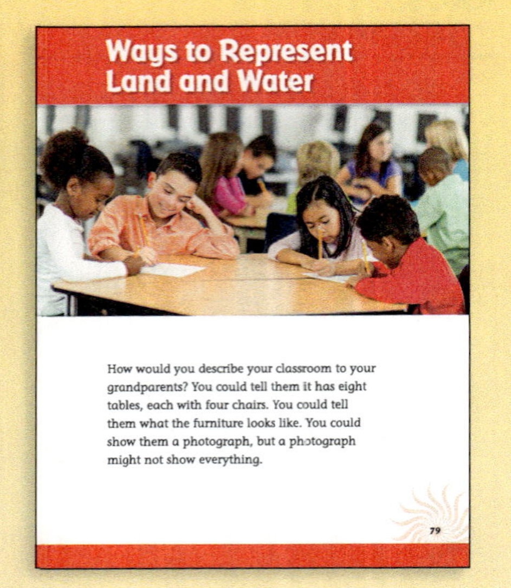

Ways to Represent Land and Water
How would you describe your classroom to your grandparents? You could tell them it has eight tables, each with four chairs. You could tell them what the furniture looks like. You could show them a photograph, but a photograph might not show everything.
79

ELA CONNECTION

These suggested strategies address the Common Core State Standards for ELA.

RI 5: Know and use text features.

SL 2: Recount or describe key ideas.

SL 3: Ask and answer questions.

SCIENCE AND ENGINEERING PRACTICES

Developing and using models

Obtaining, evaluating, and communicating information

5. Read "Ways to Represent Land and Water"

Refer students to the *Natural Sources of Water* poster and ask them to share with a partner how these bodies of water might look on a map. Explain that maps show where things are located.

Tell students that this article will explain how people can map the shapes and kinds of land and water in an area. Give students a few minutes to preview the text by looking at the photographs and illustrations. Ask students what questions they have about how people represent land and water on Earth.

Read the article aloud or have students read independently. Pause after the first section, and tell students to discuss the representations of Crater Lake with a partner. Can they locate the islands in each of the images? Mountains? How can you tell what is land and what is water on the map? What is the shape of the lake?

Read the next section and pause to let students make the connections between the photograph, the drawing, and the map of Mt. Shasta. Can they tell where the tallest peak is in all three photographs? Take turns with a partner imagining that you are at the bottom of the mountain in each of the photographs. Use your fingers to show how you would hike up to the top. What do you notice about the lines in the topographic map? Why do the colors mean?

Read the section on the Scioto River and ask students to repeat the process of making connections between the three images. What is different about this map compared to the topographic map? Can you tell where the bridge is in all three photographs? Where are the buildings?

Read the next section on aquifers. Explain that the circles and squares are fields where crops are growing. Some of the fields are in circles because that's how the sprinkler system waters the plants. The sprinkler rotates in a circle to reach all of the plants.

➤ *What do you notice about this map of the aquifer?*

➤ *How is it different than the other maps?* [Shows water that is underground.]

➤ *Why is this type of map important for farmers?*

Read the last section.

Ask students to look at the image of Earth and identify where the water is and where land is. Ask them to discuss why they think there are different colors and shades of color. Is the United States all one color? Why not? Refer students to the map of the United States and discuss what they notice with their partner. Where are the highways that connect the states? What bodies of water can they identify? How do the shapes of the images help them figure out what is a lake, river, or the ocean?

6. Map the schoolyard

This is good opportunity to have students go outside and make a simple map of the schoolyard or nearby park. Have them include sources of water.

7. Assess progress: notebook entries

Collect students' notebooks after class and assess progress.

What to Look For

- *Students are able to describe or draw one way in which engineers prevent erosion.*
- *Students drawings show that they understand how maps represent various land and water features.*

BREAKPOINT

CROSSCUTTING CONCEPTS

Scale, proportion, and quantity

ELA CONNECTION

This suggested strategy addresses the Common Core State Standards for ELA.

SL 5: Add drawings or other visual displays to recounts of experiences.

8. Review the guiding question (optional)

The day before the I-Check, you can review the guiding question.

➤ *How can we apply what we know about the ways land and water interact?*

If necessary, bring out some of the tools students used to explore the interaction. They found out what was in soil by putting samples in vials with water and shaking the mixture. The result was the separation of different particle sizes (gravel, sand, silt, and clay) with humus floating on top.

The other way students applied their knowledge was to review the engineering designs to reduce erosion. Growing plants are a good way to reduce erosion of sand and soil. Manufactured materials from rock particles (bricks, concrete shapes) and bags of sand can also serve to reduce erosion of natural earth materials.

BREAKPOINT

9. Assess progress: I-Check

When students have completed the investigation, give them *Investigation 4 I-Check.*

Review student responses. Use the What to Look For information in the Assessment chapter for guidance. Note concepts that you might want to revisit with students, using the next-step suggestions.

The students' experiences in this investigation contribute to their understanding that some events on Earth happen quickly and others occur over a very long period of time, that wind and water can change the shape of the land, that water is found in many locations on Earth and in different forms (solid ice and liquid form), and that there are different ways to represent landforms and water on Earth, such as maps. The experiences also contribute to students' understanding that the engineering design process involves defining problems, developing possible solutions, and comparing solutions to optimize the design solution.

DISCIPLINARY CORE IDEAS

ESS1.C: The history of planet Earth

ESS2.A: Earth materials and systems

ESS2.B: Plate tectonics and large-scale system interactions

ESS2.C: The roles of water in Earth's surface processes

ETS1.A: Defining and delimiting engineering problems

ETS1.B: Developing possible solutions

ETS1.C: Optimizing the design solution

INTERDISCIPLINARY EXTENSIONS

Language Extensions

- **Compare soil habitats**
  Ask if the soil on a beach is the same as the soil in students' backyards. How about the soil in the desert? Ask students to explain the differences. How would these differences affect the plant and animal life? Using the table below, compare different habitats in soil from different environments—the ocean beach, the desert, and your local soil.

	What's in the soil?	Environment	Plants and animals
Ocean beach	sand	wet or dry, depending on tides; can be sunny or cool	mostly grasses crabs
Desert	sand, gravel,pebbles	dry, hot, little water	cacti, grasses
Local			

- **Draw soil profiles**
  Have students look at the pictures on pages 46 and 47 of *Science Resources*. Explain that these pictures are "cutaways" of soil. They show what soil looks like after cutting away a slice of the soil. Ask students to draw a cutaway picture of soil that is a perfect habitat for plants and animals. Students should show the top of the ground, the layers of soil beneath the ground, the types of rocks in the soil, any roots growing in the soil, and animals that may be living in the soil. Encourage students to label the elements in their drawings. Students may also write or dictate a phrase describing how and why their soil is the perfect soil.

- **Write directions for making soil**
  Have students write down how they made soil and how they took it apart. You could prepare a set of fill-in sentences for them to complete.

Math Extension

- **Math problem**

  Sara, Maria, Lamar, Pedro, and Ann went on a rock hunt. The girls each found five rocks. Lamar found six rocks, and Pedro found four rocks.

  For each problem, show how you found your answer and write an equation.

  1. When Pedro and Lamar combined their rocks, how many rocks did they have?
  2. The girls put all their rocks together. How many rocks do they have?
  3. Pedro and Sara collected all the same-size rocks. How many rocks would they have if they put their rocks together?
  4. Ann and Pedro combined their rocks. Then Sara and Lamar combined their rocks. Which pair of students has more rocks? How do you know?
  5. Maria met her friend Arlo. Arlo collected rocks too. When Maria and Arlo added their rocks together, they had 13 rocks. How many rocks did Arlo have in his collection?
  6. Lamar wants to have a dozen rocks in his collection. He found six on the rock hunt. How many more does he need to find?

Name ______________________ Date ____________

MATH EXTENSION

Investigation 4: Soil and Water

Three girls (Sara, Maria, Ann) and two boys (Lamar and Pedro) went on a rock hunt. The girls each found five rocks. Lamar found six rocks, and Pedro found four rocks.

For each problem, show how you found your answer and write an equation.

1. When Pedro and Lamar combined their rocks, how many rocks did they have?
2. The girls put all their rocks together. How many rocks do they have?
3. Pedro and Sara collected all the same-size rocks. How many rocks would they have if they put their rocks together?
4. Ann and Pedro combined their rocks. Then Sara and Lamar combined their rocks. Which pair of students has more rocks? How do you know?
5. Maria met her friend Arlo. Arlo collected rocks too. When Maria and Arlo added their rocks together, they had 13 rocks. How many rocks did Arlo have in his collection?
6. Lamar wants to have a dozen rocks in his collection. He found six on the rock hunt. How many more does he need to find?

FOSS Pebbles, Sand, and Silt Module
Investigation 4: Soil and Water
No. 17—Teacher Master

No. 17—Teacher Master

Notes on the problem. The context of finding rocks provides an opportunity for students to solve story problems. Be sure students know the names of the girls (Sara, Maria, and Ann) and the names of the boys (Lamar and Pedro). The problems increase in difficulty. Provide concrete materials to represent the rocks for students as needed.

After students have an opportunity to solve problems independently, give them time to work with a partner. As a class, discuss the problems and solutions. Have students present their solutions. Note the different strategies for arriving at the answer.

1. This is a straightforward addition problem where two known quantities are added together. Students may use known facts, pictures, or concrete objects to determine the sum.

   The equation is $4 + 6 = 10$ (or $6 + 4 = 10$).

2. This is another straightforward addition problem, with three quantities. Each of the quantities is the same. Students may solve it with skip counting or the strategies listed above.

   The equation is $5 + 5 + 5 = 15$.

3. This is another simple addition of two addends.

   The equation is $4 + 5 = 9$ (or $5 + 4 = 9$).

4. This is a two-step problem, as it involves comparison of two sums.

   Ann and Pedro have $5 + 4 = 9$ rocks.

   Sara and Lamar have $5 + 6 = 11$ rocks.

   By direct comparison, Sara and Lamar have two more rocks.

   For students who have difficulty seeing this, make a concrete graph of objects to represent each sum and then count on two more rocks.

   - Some students may realize that they know how many rocks Ann and Pedro have, since they just determined that Pedro and Sara have 9 rocks and the girls each collected the same number of rocks.
   - Some students may be able to see that because the girls have the same number of rocks, they only need to compare the rocks the boys have.

5. In this problem two quantities are being combined, and one quantity is unknown. Maria has 5 rocks plus the number that Arlo has equals 13.

   The equation for this problem is: $5 + ? = 13$ ($? = 8$)

6. Students need to know that one dozen is equal to 12. This problem also involves a missing addend or an unknown.

   The equation for this problem is $6 + ? = 12$ ($? = 6$)

TEACHING NOTE

This is a missing-addend problem and lays the foundation for finding unknowns. Tell students they are doing algebra when they solve this problem!

Science Extension

- **Set up a soil magnification station**

  Put one vial of homemade soil in a zip bag. Put a vial of one of the soil samples in a zip bag. Have students use hand lenses to look at and compare the two bags of soil.

TEACHING NOTE

Encourage students to use the Science and Engineering Careers Database on FOSSweb.

Name ____________ Date ________

HOME/SCHOOL CONNECTION

Investigation 4: Soil and Water

Our study of rock sizes led us to a study of soil. We now know that soil is mostly rock particles and some humus.

One important property of soil is its water content. Soil scientists often do tests to see how quickly water soaks into and passes through the soil. This is called permeability. Test and compare the permeability of soil in two or more places around where you live.

Use a trowel or metal spoon to dig a shallow hole in the soil, maybe the size of a soda can, but not very deep. Pour in about a cup of water and time how long the water takes to completely soak into the soil. Compare flower beds, gardens, edges of lawns, paths, sandboxes, and so on. Keep track of the time needed to soak into different soils and collect a little sample of the soil.

Remember, in order to compare, the holes should be the same size, and the amount of water should always be the same.

Record your results. Write about what you find out.

No. 18—Teacher Master

Environmental Literacy Extensions

- **Plant seeds in sand and soil**
  After reading the *FOSS Science Resources* article called *Testing Soil,* set up a class experiment for growing plants in sand and soil. Have students find out how radish seeds grow in two different earth materials, sand and soil. Provide 150 mL (5-oz.) plastic or paper cups, radish seeds, sand, and soil. Use a pencil to poke a small drain hole in each cup. Have each pair of students fill one cup with soil and one with sand. Have them plant 4–6 radish seeds in each cup and add one vial of water to each cup. Store the cups in a basin placed in a sunny window. Have students add water every other day, if necessary. Observe what happens.

- **See what grows**
  Put into a basin the remaining soil samples that were collected on the field trip. Sprinkle the soil with some water and set it in a sunny window. After several days or weeks, grass and weed seeds that are in the soil may sprout and grow.

- **Make an earthworm terrarium**
  Earthworms are important animals that help create and enrich soil. Have students find out more about earthworms and what they need to live. Ask them to go out and find some earthworms (especially the day after a rain) and help them create a suitable classroom habitat for the worms.

- **Take a field trip to a natural source of water**
  If there is a creek or pond near the school, take a field trip to this local source of water. Or go on a longer field trip to a lake or an ocean beach.

Home/School Connection

The speed with which water can drain through soil is a property that changes with the composition of soil. At home, students become "soil engineers" as they compare "perc tests" to determine water retention of soil in two locations.

Print or make copies of teacher master 18, *Home/School Connection* for Investigation 4, and send it home with students at the end of Part 2.

Assessment

THE FOSS ASSESSMENT SYSTEM *for Grades 1–2*

"Assessment is like science…To assess our students, we plan and conduct investigations about student learning and then analyze and interpret data to develop models of what students are thinking. These models allow us to predict the effect of additional teaching addressing the patterns we notice in student understanding and misunderstanding. Assessment allows us to improve our teaching practice over time, spiraling upward." (*2016 Science Framework for California Public Schools, Kindergarten through Grade 12,* chapter 9, page 3).

An important rule of thumb in educational assessment is that assessments should be designed to meet specific purposes. One size does not fit all. Formative assessments provide short-term information about learning by making students' thinking visible in order to guide instructional decisions. Summative assessments provide valid, reliable, and fair measures of students' progress over a longer period of time, at the end of a module, or the end of the year. The purpose for the assessment determines the choice of instruments that you will use.

The FOSS assessment system is designed to assess students in grades 1 and 2 in short and medium cycles. The assessment tasks allow students to demonstrate their facility with three-dimensional understanding of science, focusing in these grades on formative assessment.

Short cycle. Embedded assessment opportunities are incorporated into each part of every investigation. These assessments use student-generated artifacts, including science notebook entries, answers to focus questions, and **performance assessments**. Embedded assessments provide daily monitoring of students' learning and practices in order to help make decisions about instructional next steps. Embedded assessments such as notebook entries and answers to focus questions focus on students' conceptual development. Performance assessments focus intently on science and engineering practices and crosscutting concepts as well as disciplinary core ideas.

Contents

NOTE
For coding guides and student work samples, go to the Assessment section in *Teacher Resources* on FOSSweb.

Medium cycle. I-Check assessment opportunities occur every 1–2 weeks after an investigation is completed. Daily embedded assessments provide a quick snapshot of students' immediate learning, and I-Checks challenge students to put this learning into action in a broader context. Now students must think about the science and engineering practices, disciplinary core ideas, and crosscutting concepts they have been learning and know when, where, and how to use them. I-Checks (short for "I check my own understanding") also provide opportunities for guided self-assessment, an important skill for future learning and development of a growth mindset. Properly executed feedback can help a student focus attention on areas that need strengthening. When a student responds to feedback, you can develop an even more precise understanding of the students' learning. A feedback/response dialogue can develop into a highly differentiated path of instruction tailored precisely to the learning requirements of individual students.

ASSESSMENT *for the NGSS*

A Framework for K–12 Science Education (National Research Council, 2012), *Next Generation Science Standards* (National Academies Press, 2013), *Developing Assessment for the Next Generation Science Standards* (National Research Council, 2014), and many state frameworks provide a new vision for science education. These documents emphasize the idea that science education should resemble the way that scientists work. Students plan and conduct investigations, gather data to construct explanations, and engage in argumentation in order to build their understanding of the natural world. They apply that knowledge to engineering problems and design solutions. Students are expected to construct and discuss explanations and model systems in more and more sophisticated ways as they move through the grades. Assessment plays an important role in this new vision of science education—assessment is the bridge between teaching and learning.

"Assessment plays an important role in this new vision of science education—assessment is the bridge between teaching and learning."

Several key points in these foundational documents provide guidance for a well-designed assessment system. FOSS has followed these guidelines to ensure a robust assessment system that provides valuable diagnostic information about students' learning.

Assessment tasks should consist of multiple components in order to measure all three dimensions of science and engineering learning. The FOSS assessment system provides multiple tools and strategies to assess the three dimensions: (1) science and engineering practices, (2) disciplinary core ideas, and (3) crosscutting concepts. These tools and strategies provide evidence about what students can do, their developing conceptual understandings, and the connections that they are making among disciplines. Entry-level assessments, given before instruction begins, show what students can do and what they know before they begin a new module.

Assessment systems should include formative and summative tasks. Formative assessment tasks are embedded in the curriuculum at key stages in instruction. These tasks are designed to support teachers in collecting and analyzing data about students' conceptual understanding and growing practice. Notebook entries, answers to focus questions, response sheets, performance assessments, and oral presentations/ interviews provide the information you need to decide what students need to do next to move toward a learning goal. FOSS suggestions for next-step strategies help you address students' developing conceptions and provide information for differentiated instruction as needed.

"Our research has shown that the reflective-assessment practice provides evidence-based information that is crucial for differentiating instruction for all students—this practice can make a significant difference in students' overall achievement."

Summative assessments are designed to provide valid, reliable, and fair measures of students' progress. The FOSS system includes three types of summative assessments: *Posttests,* interim assessments, and portfolios. These assessments include multicomponent tasks including open-ended constructed-response problems, as well as some multiple-choice and short-answer items. *Posttests* are given at the end of a module and interim assessments can be given twice during a module or as an end-of-year assessment. Students can also collect work products at the end of each investigation for inclusion in a portfolio. Coding guides found on FOSSweb provide teachers with guidance when evaluating these assessments. All written assessments are consistent with grade-level writing and mathematics in the Common Core State Standards for ELA/Literacy and the CCSS and for Mathematics.

Assessment systems should support classroom instruction. The main purpose of the FOSS assessment system is to support classroom instruction—to provide the bridge between teaching and learning. Teachers need information daily about what students have learned or may be confused about. FOSS has developed a technique in which teachers spend only 10 minutes after a lesson, using a reflective-assessment practice (explained in detail later in this chapter), to gather data to determine instructional next steps. Are students ready to move on to the next lesson, or do they need some additional clarification? Our research has shown that the reflective-assessment practice provides evidence-based information that is crucial for differentiating instruction for all students—this practice can make a significant difference in students' overall achievement.

Assessment developers need to take a rigorous approach to the process of designing and validating assessments. The FOSS assessment system is based on a construct-modeling approach for assessment design. That means that we have done the research needed to describe a conceptual framework and learning performances that provide evidence of students' progressive learning (see the Framework and NGSS chapter), and we have done the technical work needed to ensure that assessment tasks provide valid, reliable, and fair evidence of students' learning. (See the Benchmark Assessment section for a more detailed description of the design behind the FOSS assessment system.)

Assessment systems should include an interpretive system and locate students along a sequence of progressively more complex understanding. FOSS provides extensive support for interpreting assessment information. For each embedded assessment, specific information in the Getting Ready and Guiding the Investigation sections describes what students do and what to look for in student responses to assess progress. Coding guides, found on FOSSweb, are provided for each item on benchmark assessments (I-Checks and the *Survey/Posttest*, as well as intermin assessments). Samples of student work, especially for open-response questions, are also available on FOSSweb. Resources in this chapter and on FOSSweb provide you with information about how to use the coding guides, as well as what to do for next steps when students need to spend more time on a practice or core idea, or to look at it from a different perspective to see more connections.

You can use FOSSmap to have students take assessments online and generate a number of diagnostic and summary reports (delivered as PDFs). Some of these reports provide information about class progress; others provide individual students and parents with information about what students know and what they still need to work on. (See the FOSSmap and Online Assessment section in this chapter.)

The FOSS assessment system was developed over a period of 5 years with data from more than 500 teachers and their students. We know that teachers can employ this assessment system for the benefit of their students, and we know that students achieve more. Perhaps even more important is the change in classroom culture that occurs when assessment is thoughtfully employed as the bridge between teaching and learning. Assessment is no longer a stress factor for students or teachers. It encourages all to adopt a growth mindset—if I know where my strengths and weaknesses are and I continue to be thoughtful and work hard, I can make progress. It models what scientists do. Scientists use the information they have to argue for the best explanation, but they keep an open mind, so that when new evidence emerges they can incorporate that into their thinking, too. That's also what good curriculum and assessment are all about.

"Assessment is no longer a stress factor for students or teachers. It encourages all to adopt a growth mindset—if I know where my strengths and weaknesses are and I continue to be thoughtful and work hard, I can make progress."

NGSS Performance Expectations

"The NGSS are standards or goals, that reflect what a student should know and be able to do; they do not dictate the manner or methods by which the standards are taught. . . . Curriculum and assessment must be developed in a way that builds students' knowledge and ability toward the PEs [performance expectations]" (*Next Generation Science Standards*, 2013, page xiv). The FOSS assessment system includes embedded, performance, and benchmark assessments. The chart displayed on this and the next page provides an overview of these assessments across the three third-grade modules. These assessments help students build knowledge and ability in concert with active investigations and readings to meet the goals of the NGSS.

Grade 2 NGSS Performance Expectations	FOSS Module	
	Embedded Assessment	Benchmark Assessment
2-PS1-1. Plan and conduct an investigation to describe and classify different kinds of materials by their observable properties.	**Solids and Liquids** • Inv 1, Part 1: notebook entry • Inv 1, Part 2: notebook entry • Inv 1, Part 3: performance assessment • Inv 1, Part 4: performance assessment • Inv 2, Part 1: performance assessment • Inv 2, Part 2: notebook entry • Inv 2, Part 3: notebook entry • Inv 3, Part 1: notebook entry • Inv 3, Part 2: performance assessment • Inv 3, Part 3: performance assessment • Inv 3, Part 4: notebook entry • Inv 4, Part 2: notebook entry • Inv 4, Part 3: performance assessment **Pebbles, Sand, and Silt** • Inv 2, Part 3: notebook entry • Inv 3, Part 2: notebook entry • Inv 3, Part 4: notebook entry	**Solids and Liquids** • *Investigation 1 I-Check* • *Investigation 2 I-Check* • *Investigation 3 I-Check* • *Investigation 4 I-Check*
2-PS1-2. Analyze data obtained from testing different materials to determine which materials have the properties that are best suited for an intended purpose.	**Solids and Liquids** • Inv 4, Part 1: notebook entry **Pebbles, Sand, and Silt** • Inv 3, Part 1: notebook entry	**Solids and Liquids** • *Investigation 1 I-Check* • *Investigation 4 I-Check*
2-PS1-3. Make observations to construct an evidence-based account of how an object made of a small set of pieces can be disassembled and made into a new object.	**Solids and Liquids** • Inv 1, Part 4: performance assessment	**Solids and Liquids** • *Investigation 1 I-Check*
2-PS1-4. Construct an argument with evidence that some changes caused by heating or cooling can be reversed and some cannot.	**Solids and Liquids** • Inv 4, Part 3: notebook entry	**Solids and Liquids** • *Investigation 4 I-Check*

Grade 2 NGSS Performance Expectations	FOSS Module	
	Embedded Assessment	Benchmark Assessment
2-LS2-1. Plan and conduct an investigation to determine if plants need sunlight and water to grow.	**Insects and Plants** • Inv 2, Part 2: performance assessment	**Insects and Plants** • *Investigation 2 I-Check*
2-LS2-2. Develop a simple model that mimics the function of an animal dispersing seeds or pollinating plants.	**Insects and Plants** • Inv 5, Part 4: performance assessment	**Insects and Plants** • *Investigation 2 I-Check* • *Investigation 5 I-Check*
2-LS4-1. Make observations of plants and animals to compare the diversity of life in different habitats.	**Insects and Plants** • Inv 1, Part 1: notebook entry • Inv 1, Part 2: performance assessment • Inv 2, Parts 1 and 3: notebook entry • Inv 2, Part 2: performance assessment • Inv 3, Parts 1–3: notebook entry • Inv 3, Part 4: performance assessment • Inv 4, Parts 1–3: notebook entry • Inv 4, Part 4: performance assessment • Inv 5, Part 1: performance assessment • Inv 5, Parts 2–3: notebook entry	**Insects and Plants** • *Investigation 1 I-Check* • *Investigation 2 I-Check* • *Investigation 3 I-Check* • *Investigation 4 I-Check* • *Investigation 5 I-Check*
2-ESS1-1. Use information from several sources to provide evidence that Earth events can occur quickly or slowly.	**Pebbles, Sand, and Silt** • Inv 1, Part 1: notebook entry • Inv 1, Part 2: notebook entry	**Pebbles, Sand, and Silt** • *Investigation 2 I-Check*
2-ESS2-1. Compare multiple solutions designed to slow or prevent wind or water from changing the shape of the land.	**Pebbles, Sand, and Silt** • Inv 4, Part 4: notebook entry	**Pebbles, Sand, and Silt** • *Investigation 4 I-Check*
2-ESS2-2. Develop a model to represent the shapes and kinds of land and bodies of water in an area.	**Pebbles, Sand, and Silt** • Inv 2, Part 2: notebook entry • Inv 4, Part 4: notebook entry	**Pebbles, Sand, and Silt** • *Investigation 4 I-Check*
2-ESS2-3. Obtain information to identify where water is found on Earth and that it can be solid or liquid.	**Pebbles, Sand, and Silt** • Inv 4, Part 3: notebook entry	**Pebbles, Sand, and Silt** • *Investigation 4 I-Check*
K-2-ETS1-1. Ask questions, make observations, and gather information about a situation people want to change to define a simple problem that can be solved through the development of a new or improved object or tool.	**Pebbles, Sand, and Silt** • Inv 2, Part 1: performance assessment • Inv 4, Part 1: performance assessment	
K-2-ETS1-2. Develop a simple sketch, drawing, or physical model to illustrate how the shape of an object helps it function as needed to solve a given problem.	**Solids and Liquids** • Inv 1, Part 4: performance assessment **Insects and Plants** • Inv 2, Part 2: performance assessment	
K-2-ETS1-3. Analyze data from tests of two objects designed to solve the same problem to compare the strengths and weaknesses of how each performs.	**Pebbles, Sand, and Silt** • Inv 3, Part 3: performance assessment	

EMBEDDED *Assessment*

Assessment is the bridge that connects teaching and learning. Assessing students on a regular basis gives you valuable information that guides instruction and keeps families and other interested members of the educational community informed about students' progress.

Embedded assessments are suggested for most investigation parts. You will find a description of what and when to assess in the **Getting Ready** section of each part of each investigation. For example, here is the Getting Ready assessment step for Part 1 of the first investigation.

18. Plan assessment: notebook entry

In Step 13 of Guiding the Investigation, students complete their first notebook entry (notebook sheet 1, *Rubbing Rocks*). Depending on when you are teaching this module in the school year, you may want to scaffold this entry for the class. Review students' work to make sure they know that smaller roks come from bigger rocks (when they are rubbed together). (This is one of the processes that contributes to the slow but constant changes to Earth's surface.)

As you progress through the lesson and it is time to assess, you will find a step in **Guiding the Investigation** that reminds you about what to assess. It provides a bulleted list of what to look for when you review the students' work for disciplinary core ideas, science and engineering practices, and crosscutting concepts.

14. Assess progress: notebook entry

Collect students' notebook after class and review their work.

What to Look For

- *Students recorded their observations by completing the picture to show small pieces of rock falling from the rocks to the table.*
- *The effect of rubbing rocks together is that "tiny pieces break off" or "makes sand."*
- *When rocks rub together, it can change the shape of some rocks quickly; others can change more slowly.*

Performance Assessments

Assessing the three dimensions envisioned in the NRC *Framework* and the NGSS performance expectations challenges students to be engaged in science and engineering practices in order to build disciplinary core ideas bridged by crosscutting concepts. This is an everyday occurence in the FOSS curriculum. One part in each investigation has been designated as a performance assessment for you to formatively check students' progress for all three dimensions at the same time. You peek over students' shoulders while they are in the act of doing science or engineering. You take note of what they are doing and discussing and sometimes conduct short interviews. Observing the rich conversation among students and the actions they are taking to investigate phenomena or design solutions to problems provides important information about student progress. At times, you might step in with a 30-second interview to ask a few carefully crafted questions to learn more about students' deeper conceptual understanding and practices. The What to Look For bullets in each performance assessment step will help you focus on pertinent science and engineering practices, disciplinary core ideas, and crosscutting concepts. You can record student progress on the *Performance Assessment Checklist* for each part.

Performance Assessment Checklist—Pebbles, Sand, and Silt

Start date ________
End date ________

	Investigation 1, Part 3				
Student names	Planning and carrying out investigations	Constructing explanations	Engaging in argument from evidence	PS1.A: Structure and properties of matter	Patterns

Science Notebook Entries

Making good observations and using them to develop explanations about the natural world is the essence of science. This process calls for critical thinking and honed communication skills. Science notebook entries are designed specifically to help you understand the practices, crosscutting concepts, and scientific explanations and knowledge that students are developing.

Three kinds of notebook entries serve as assessments for learning. Each part of each investigation is driven by a **focus question**. Each part usually concludes with students writing an answer to the focus question in their notebooks. Their answers reveal how well they have made sense of the investigation and whether they have focused on the relevant actions and discussions. Prepared **notebook sheets** or **free-form notebook entries** provide information about how students make and organize observations and how they think about analyzing and interpreting data.

NOTE
You only need 10 minutes after a lesson to review student work and gather evidence of learning. See the reflective-assessment practice on the next page.

Using the Reflective-Assessment Practice

Successful teachers incorporate a system of continuous formative assessment into their standard teaching practices. One of the keys to formative assessment is frequency. The more often you can gather evidence about students' progress, the more able you will be to guide each student's path to understanding. The **reflective-assessment practice** is a proven method for gathering that information. It takes little time but has a big impact on students' learning.

Reflective-Assessment Practice

1. Anticipate
Use the Investigations Guide to plan for each part and determine embedded assessment.

2. Teach
Use Guiding the Investigation to teach the lesson. Collect student notebooks.

3. Review
Review students work (10 minutes). Use "What to Look For" in Guiding the Investigation.

4. Reflect
Note trends and patterns you see in student understanding.

5. Adjust
Plan next instructional steps based on assessment reflection. Make notes for next year.

AFTER EACH PART

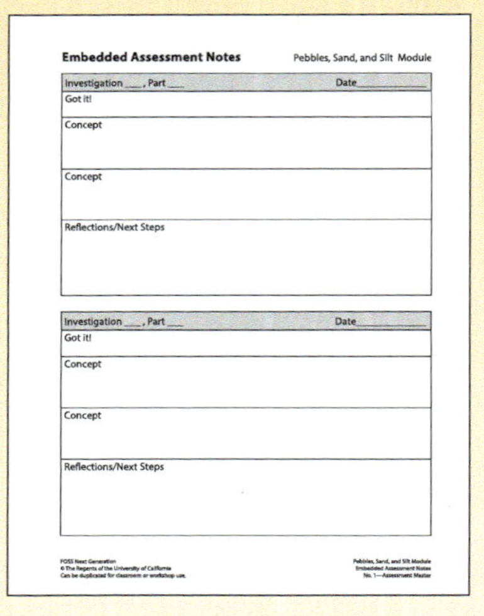
Embedded Assessment Notes Pebbles, Sand, and Silt Module

Investigation ___, Part ___	Date___
Got it!	
Concept	
Concept	
Reflections/Next Steps	

Investigation ___, Part ___	Date___
Got it!	
Concept	
Concept	
Reflections/Next Steps	

You can record notes for two lessons on each page.

Print or make copies of ***Embedded Assessment Notes*** to record students' progress with the embedded assessments in each part of an investigation (see the What to Look For bullets). If you don't think you can review every student's work in 10 minutes, choose a random sample. Many teachers are pleasantly surprised to discover how many students they can review in that short amount of time. The important thing is that you are looking at student work on a daily basis as often as possible and looking for patterns that reveal strengths and areas that need additional support.

FOSS

1. Anticipate. Check the Getting Ready section for the suggested embedded assessment for the part you are planning. Before class, fill in the investigation and part number along with the date on *Embedded Assessment Notes*. Check the assessment step in Guiding the Investigation, and fill in the things you are looking for in student thinking. Limit your assessment to one or two important ideas.

2. Teach. Follow the steps in Guiding the Investigation.

3. Review. Collect the notebooks at the end of class. Have students turn in their notebooks *open to the page you will be reviewing.* (This may sound trivial, but it will save you a lot of time.) If you didn't write in your "What to look fors" before teaching the lesson, do that now by checking the assessment step in Guiding the Investigation. Use *Embedded Assessment Notes* to record what you observe. Make a tally mark for each student who "got it"; write in names and notes for students who need help. Spend no more than 10 minutes on this review.

TEACHING NOTE

We encourage you to stop after 10 minutes even though you may be tempted to spend more time. If you keep this process quick and easy you are more likely to use it frequently. Frequency is key to the success of this practice.

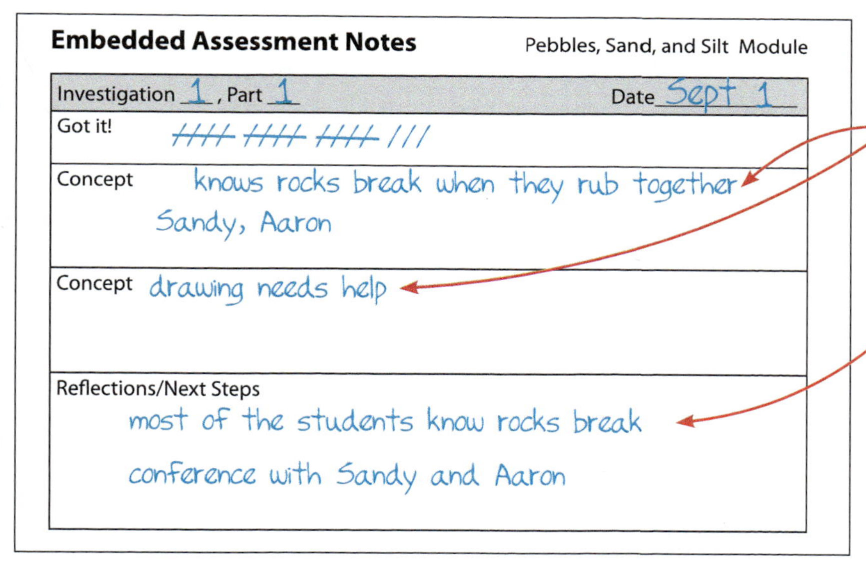
Embedded Assessment Notes Pebbles, Sand, and Silt Module

Investigation 1, Part 1	Date Sept 1
Got it!	~~||||~~ ~~||||~~ ~~||||~~ |||
Concept	knows rocks break when they rub together Sandy, Aaron
Concept	drawing needs help
Reflections/Next Steps	most of the students know rocks break conference with Sandy and Aaron

Note what to look for and make notes as you review student work
Write reflections and next steps here after 10 minutes of review.

4. Reflect. Take 5 minutes to summarize the trends and patterns (highlights and challenges) you saw, and record notes in the "Reflections/Next Steps" section.

5. Adjust. In the same section of the sheet, describe the next steps you will take to clarify any problems, or note highlights you saw in students' progress. Some suggestions appear later in this chapter and accompany the coding guides. This is the defining factor in formative assessment. You must take some action to help students improve. If you do this process frequently, the next steps required should take only a few minutes of class time when the next part begins.

BENCHMARK *Assessment*

I-Checks

At the end of each investigation, students take an I-Check benchmark assessment. I-Checks (short for "I check my own understanding") provide students with an opportunity to demonstrate their learning. When you return the I-Checks, students have yet another opportunity to think about their understanding. I-Checks can also provide you with insight into how well students are beginning to form a set of scientific "rules" to take out into the real world.

Students in primary grades, as well as English learners, may experience difficulty reading questions on assessments. For this reason, we suggest that you read the questions aloud. Specific directions for administering each question are provided on the following pages.

After students complete the I-Check, determine their progress by reviewing each item, using the information provided under the heading "What to Look For" in this chapter. We recommend that you review one item at a time; that is, review item 1 for all students, then move on to item 2, and so on. Even though you have to shuffle papers more, you will find that it takes less time overall to review the assessments. Proceeding one item at a time, rather than one student at a time, allows you to establish a mind-set for each item, think about the whole class's performance on an item, and determine the next steps you might take.

I-Checks are most valuable when you use them as formative assessments. Research shows that students learn more when they take part in evaluating their own responses. To do this, review students' tests, record their progress on the *Assessment Record* sheet, but *do not write on them*. Thinking about the questions comes to a halt when students see marks on their papers made by their teachers. Using class discussion and colored pencils, help students confirm (underline in green), complete (add more to the response in blue), and correct (cross out and edit in red) their answers to clarify and refine their thinking. You can invent many ways to help students reflect on their work and make this an inviting practice. The important thing is that students make the marks on their papers because they searched their own thinking and are compelled to update their responses.

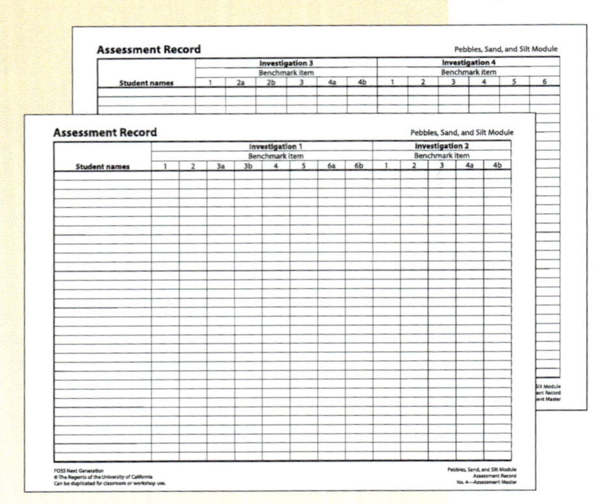

Assessment Record — Pebbles, Sand, and Silt Module

Student names	Investigation 3 Benchmark item						Investigation 4 Benchmark item					
	1	2a	2b	3	4a	4b	1	2	3	4	5	6

Assessment Record — Pebbles, Sand, and Silt Module

Student names	Investigation 1 Benchmark item								Investigation 2 Benchmark item				
	1	2	3a	3b	4	5	6a	6b	1	2	3	4a	4b

FOSS Next Generation
© The Regents of the University of California
Can be duplicated for classroom or workshop use.

Pebbles, Sand, and Silt Module
Assessment Record
No. 4—Assessment Master

Using the What-to-Look-For Guides

This module has four investigations, so it has four I-Checks. Each I-Check is two or three pages long, with four to five items per I-Check. In this chapter, each page of an I-Check is represented by a two-page spread (see reduction below)—the left-hand page explains how to administer the I-Check and provides pointers regarding what to look for in students' responses. The right-hand page shows a sample of possible student work.

If you find that students have not progressed to a satisfactory level of performance on an item, look at the next-steps suggestion for that item.

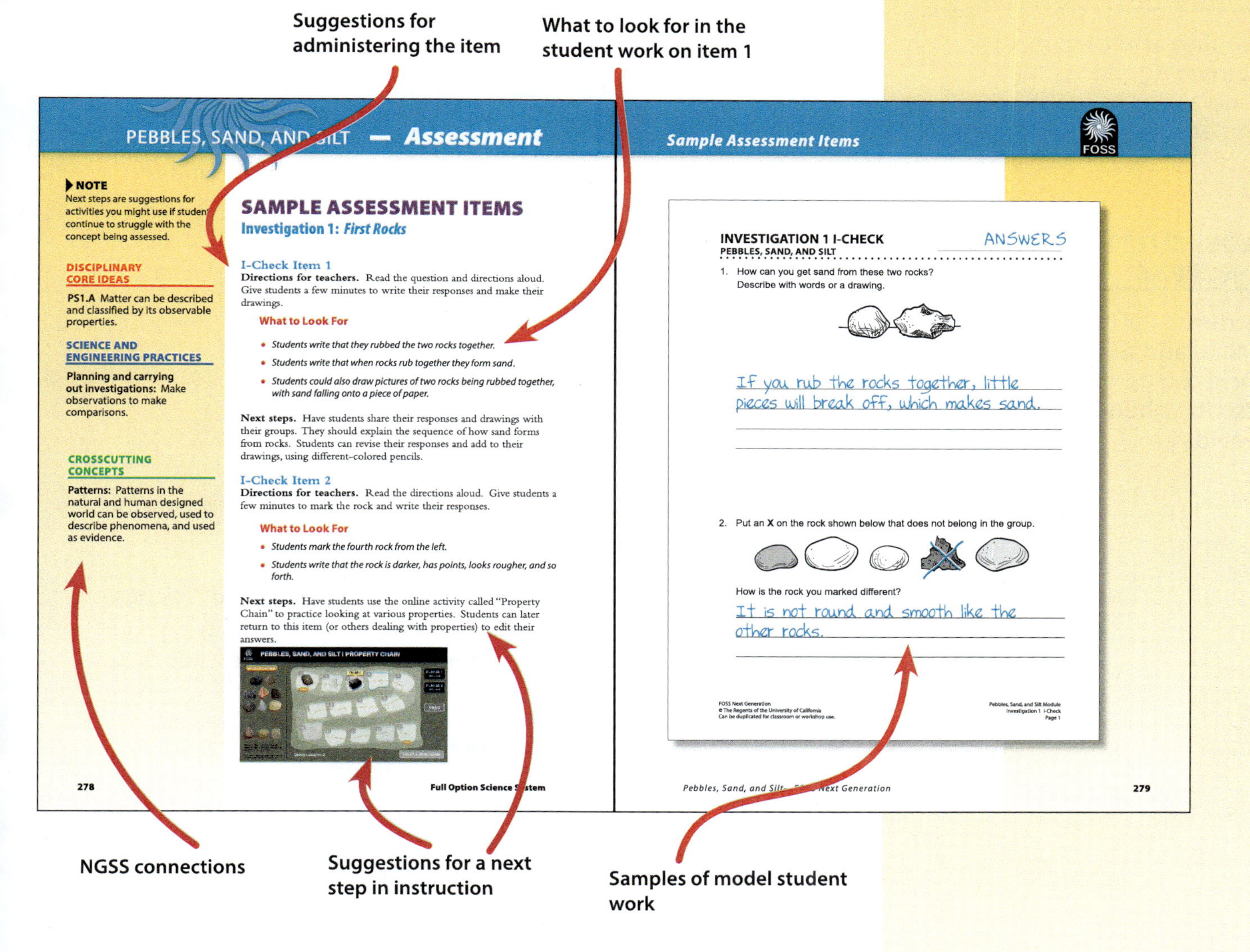

NOTE
Next steps are suggestions for activities you might use if students continue to struggle with the concept being assessed.

DISCIPLINARY CORE IDEAS

PS1.A Matter can be described and classified by its observable properties.

SCIENCE AND ENGINEERING PRACTICES

Planning and carrying out investigations: Make observations to make comparisons.

CROSSCUTTING CONCEPTS

Patterns: Patterns in the natural and human designed world can be observed, used to describe phenomena, and used as evidence.

SAMPLE ASSESSMENT ITEMS

Investigation 1: *First Rocks*

I-Check Item 1

Directions for teachers. Read the question and directions aloud. Give students a few minutes to write their responses and make their drawings.

What to Look For

- *Students write that they rubbed the two rocks together.*
- *Students write that when rocks rub together they form sand.*
- *Students could also draw pictures of two rocks being rubbed together, with sand falling onto a piece of paper.*

Next steps. Have students share their responses and drawings with their groups. They should explain the sequence of how sand forms from rocks. Students can revise their responses and add to their drawings, using different-colored pencils.

I-Check Item 2

Directions for teachers. Read the directions aloud. Give students a few minutes to mark the rock and write their responses.

What to Look For

- *Students mark the fourth rock from the left.*
- *Students write that the rock is darker, has points, looks rougher, and so forth.*

Next steps. Have students use the online activity called "Property Chain" to practice looking at various properties. Students can later return to this item (or others dealing with properties) to edit their answers.

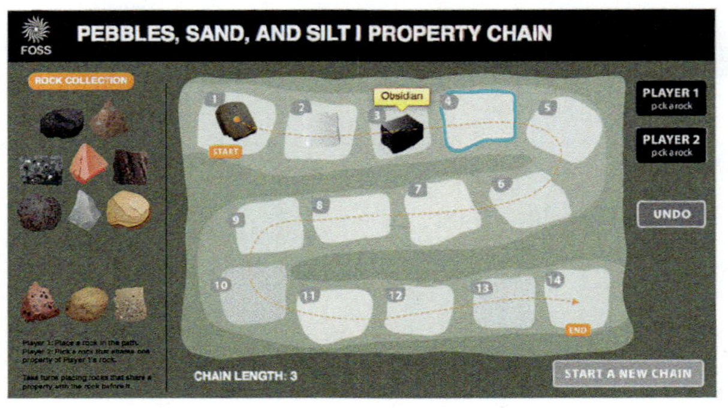

INVESTIGATION 1 I-CHECK

PEBBLES, SAND, AND SILT

ANSWERS

1. How can you get sand from these two rocks?
   Describe with words or a drawing.

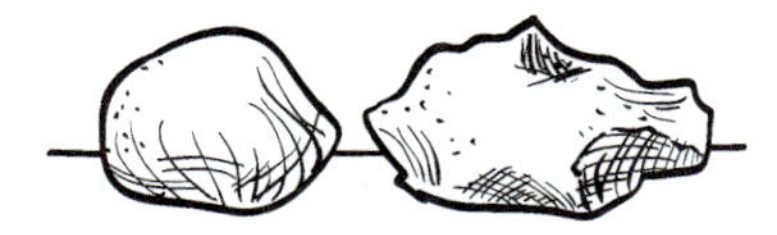

If you rub the rocks together, little pieces will break off, which makes sand.

2. Put an **X** on the rock shown below that does not belong in the group.

    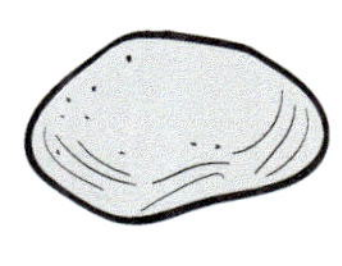

How is the rock you marked different?

It is not round and smooth like the other rocks.

FOSS Next Generation

Pebbles, Sand, and Silt Module
Investigation 1 I-Check
Page 1

DISCIPLINARY CORE IDEAS

PS1.A Matter can be described and classified by its observable properties.

DISCIPLINARY CORE IDEAS

ESS1.C Some events happen very quickly; others occur very slowly, over a time period much longer than one can observe.

CROSSCUTTING CONCEPTS

Cause and effect: Events have causes that generate observable patterns.

I-Check Item 3ab

Directions for teachers. For item 3a, give a scoria rock to each student. Read the directions aloud. Give students a few minutes to write their responses.

For item 3b, have students keep the scoria rock. Read the directions aloud. Confirm that the drawing shows water in a cup. Have students draw the rock after it has been placed in the water and describe what they would see. Give students a few minutes to write and draw.

What to Look For

- *Students describe two properties, e.g., the color of the rock (red, black, gray, dark); the texture (bumpy, rough, has holes); when you pick up the rock, it feels light (not as heavy as expected) (3a).*
- *For item 3b, students draw and describe the rock at the bottom of the cup with bubbles on the rock and coming off the rock; and the rock in water darker than the rock out of water (3b).*

Next steps. Have students share their responses and drawings with their groups. After discussing and editing properties (item 3a), give each group a cup of water. Let them drop their scoria rocks in the water, then add to their responses and drawings using a different-colored pencil for item 3b.

I-Check Item 4

Directions for teachers. Read the statement and directions aloud. Read aloud the three words. Give students time to circle the best word to complete the sentence.

What to Look For

- *Students circle "weathering" as the process of rocks breaking into smaller pieces.*

Next steps. Have each group discuss the definition of each word. Next, tell students you are going to describe the meaning of one of the words. They should hold up one, two, or three fingers to indicate which word is being described. *Sinking* is 1, *weathering* is 2, and *washing* is 3. Here are possible definitions for the three words.

1. *When rocks are placed in a cup of water they drop to the bottom, this is called _____ . [Sinking.]*
2. *When rocks are covered with dirt, you place them in water to get the rocks clean. This is called _____ . [Washing.]*
3. *When rocks rub together and break into smaller pieces, this is called _____ . [Weathering.]*

INVESTIGATION 1 I-CHECK

PEBBLES, SAND, AND SILT

ANSWERS

3. Look at the scoria rock your teacher gave you.
   a. Describe two of its properties.

The rock is red.
The rock is bumpy.
The rock is not heavy.

   b. Draw and describe what you might see when this scoria rock is put into water.

The rock sinks.
There are bubbles.
The rock is a darker color.

4. When rocks rub together and break into smaller pieces, this is called _____ .
   (Circle the one best answer.)

sinking  weathering washing

Pebbles, Sand, and Silt Module
Investigation 1 I-Check
Page 2

TEACHER NOTES

TEACHER NOTES

TEACHER NOTES

TEACHER NOTES

TEACHER NOTES

TEACHER NOTES